Real-Time PCR

An Essential Guide

Edited by:

Kirstin Edwards, Julie Logan and Nick Saunders

Genomics Proteomics and Bioinformatics Unit,
Specialist and Reference Microbiology Division,
Health Protection Agency,
London,
UK

Copyright © 2004
Horizon Bioscience
32 Hewitts Lane
Wymondham
Norfolk NR18 0JA
U.K.

www.horizonbioscience.com

British Library Cataloguing-in-Publication Data

A catalogue record for this book is available from the British Library

ISBN: 0-9545232-7-X

Printed and bound in Great Britain

Contents

Contributors

Tom Brown
University of Southampton
Department of Chemistry
Highfield
Southampton SO17 1BJ
UK

Stephen A Bustin*
Centre for Academic Surgery
4th Floor Alexandra Wing
The Royal London Hospital
London E1 1BB
UK
s.a.bustin@qmul.ac.uk

Kirstin J. Edwards*
Genomics Proteomics and Bioinformatics Unit
Specialist and Reference Microbiology Division
Health Protection Agency
61 Colindale Avenue
London NW9 5HT
UK
kirstin.edwards@hpa.org.uk

Julie D. Fox*
University of Calgary and Provincial
Laboratory for Public Health (Microbiology)
3030 Hospital Drive NW
Calgary, Alberta T2N 4W4
Canada
J.Fox@provlab.ab.ca

Sam Hibbitts
Department of Medical Microbiology
University of Wales College of Medicine
Heath Park
Cardiff CF14 4XN
UK

Nilgun Isik
University of Istanbul
Istanbul Medical Faculty
Dept. Microbiology and Clinical Microbiology
34390 Çapa
Istanbul
Turkey

Martin A. Lee*
Dstl Porton Down
Salisbury
Wiltshire SP4 0JQ
UK
martinalanlee@aol.com

Dario L. Leslie
Dstl Porton Down
Salisbury
Wiltshire SP4 0JQ
UK

Julie M.J. Logan*
Genomics Proteomics and Bioinformatics Unit
Specialist and Reference Microbiology Division
Health Protection Agency
61 Colindale Avenue
London NW9 5HT
UK
julie.logan@hpa.org.uk

Tania Nolan
Stratagene Inc
La Jolla
CA
USA

Parvinder Punia
Genomics Proteomics and Bioinformatics Unit
Specialist and Reference Microbiology Division
Health Protection Agency
61 Colindale Avenue
London NW9 5HT
UK

Andrew D. Sails*
Health Protection Agency
Institute of Pathology
Newcastle General Hospital
Newcastle upon Tyne NE4 6BE
UK
andrew.sails@hpa.org.uk

Nick A. Saunders*
Genomics Proteomics and Bioinformatics Unit
Specialist and Reference Microbiology Division
Health Protection Agency
61 Colindale Avenue
London NW9 5HT
UK
nick.saunders@hpa.org.uk

David J. Squirrell
Dstl Porton Down
Salisbury
Wiltshire SP4 0JQ
UK

* Corresponding author

Preface

The polymerase chain reaction (PCR) is a technique so commonplace in the modern day laboratory that it is easy to forget its revolutionary impact. Real-time PCR has removed many of the limitations of standard end-point PCR and since its introduction in the mid-1990s there has been an explosion both in the number of publications and available instrumentation describing real-time PCR applications across many disciplines.

This book aims to provide both the novice and experienced user with an invaluable point of reference to the technology, instrumentation and its wide range of applications. The initial chapters cover the important aspects of real-time PCR, from choosing an instrument and probe system to set-up, controls and validation. It then goes on to give a comprehensive overview of important real-time applications such as quantitation, expression analysis and mutation detection. This is complemented by the final chapters which address the diagnosis of infectious diseases. This essential manual should serve both as a basic introduction to real-time PCR and a source of current trends and applications for those already familiar with the technology. We also hope this text will stimulate readers of all levels to develop their own innovative approaches to real-time PCR.

Julie M. J. Logan
Nick A. Saunders
Kirstin J. Edwards

Books of Related Interest

Full details of all these books at: www.horizonbioscience.com

1

An Introduction
to Real-Time PCR

N. A. Saunders

Abstract

The development of instruments that allowed real-time monitoring of fluorescence within PCR reaction vessels was a very significant advance. The technology is very flexible and many alternative instruments and fluorescent probe systems have been developed and are currently available. Real-time PCR assays can be completed very rapidly since no manipulations are required post-amplification. Identification of the amplification products by probe detection in real-time is highly accurate compared with size analysis on gels. Analysis of the progress of the reaction allows accurate quantification of the target sequence over a very wide dynamic range, provided suitable standards are available. Further investigation of the real-time PCR products within the original reaction mixture using probes and melting analysis can detect sequence variants including single base mutations.

Since the first practical demonstration of the concept real-time PCR has found applications in many branches of biological science. Applications include gene expression analysis, the diagnosis of infectious disease and human genetic testing.

Due to their capability in fluorimetry the real-time machines are also compatible with alternative amplification methods such as NASBA provided a fluorescence end-point is available.

PCR: The Early Years

The theoretical concept of producing many copies of a specific DNA molecule by a cycling process using DNA polymerase and oligonucleotide primers was first expounded in a paper by Kleppe and colleagues in 1971(Kleppe *et al.*, 1971). At that time, the practical exploitation of such a process must have seemed remote to the biologists who read the paper. This was due to the difficulty and cost of producing oligonucleotides, the non-availability of thermostable DNA polymerases and the lack of automated thermocycling instruments. By the time of the first demonstration of the PCR process by Saiki and colleagues in 1985 (Saiki *et al.*, 1985) automated oligonucleotide synthesisers were commonly available. This meant that the potential of PCR in a wide range of applications was recognised. However, it was still necessary to inject fresh thermo-labile polymerase prior to each elongation step and thermal cyclers were still in development. Consequently, the key step in realising the potential of the PCR was probably the use of a thermostable polymerase which was first described in 1988 (Saiki *et al.*, 1988). Since the first description of a practical DNA amplification process many refinements have been described and automatic thermal cyclers have become standard laboratory equipment. PCR is now an essential tool for many biologists and the standard protocols are very simple and user friendly. The exponential amplification process provides nanogram quantities of essentially identical DNA molecules starting from a few copies of a target sequence. The amplified material (the PCR amplicon) is available in sufficient quantity to be identified by size analysis, sequencing or by probe hybridisation. It can also be cloned readily or used as a reagent.

The Need for Real-Time PCR

Much of the technical effort involved in standard PCR is now directed toward positive recognition of the amplicons. The important methods of post-PCR analysis rely on either the size or sequence of the amplicon. Gel electrophoresis is often used to measure the size of the amplicon and this is both inexpensive and simple to implement. Unfortunately, size analysis has limited specificity since different molecules of approximately the same molecular weight cannot be distinguished. Consequently, gel electrophoresis alone is not a sufficient PCR end-point in many instances, including most clinical applications. Characterisation of the product by its sequence is far more reliable and informative. Probe hybridisation assays for this purpose are available but many are multi-step procedures. Such methods are time-consuming and care must be taken to ensure that amplicons accidentally released into the laboratory environment do not contaminate the DNA preparation and clean rooms.

Real-time PCR machines greatly simplify amplicon recognition by providing the means to monitor the accumulation of specific products continuously during cycling. All current instruments designed for real-time PCR measure the progress of amplification by monitoring changes in fluorescence within the PCR tube. Changes in fluorescence can be linked to product accumulation by a variety of methods. A further advantage of the real-time format is that the analysis can be performed without opening the tube which can then be disposed of without the risk of dissemination of PCR amplicons or other target molecules into the laboratory environment. Although alternative methods for avoiding PCR contamination are available, containment within the PCR vessel is likely to be the most efficient and cost-effective. A major drawback of standard PCR formats that rely on end-point analysis is that they are not quantitative because the final yield of product is not primarily dependent upon the concentration of the target sequence in the sample. Real-time PCR overcomes this limitation.

3

Real-Time PCR Chemistries

There are two general approaches used to obtain a fluorescent signal from the synthesis of product in PCR. The first depends upon the property of fluorescent dyes such as SYBR Green I to bind to double stranded DNA and undergo a conformational change that result in an increase in their fluorescence. The second approach is to use fluorescent resonance energy transfer (FRET). These methods use a variety of means to alter the relative spatial arrangement of photon donor and acceptor molecules. These molecules are attached to probes, primers or the PCR product and are usually selected so that amplification of a specific DNA sequence brings about an increase in fluorescence at a particular wavelength.

A major advantage of the real-time PCR instruments and signal transduction systems currently available is that it is possible to characterise the PCR amplicon *in situ* on the machine. This is done by analysis of the melting temperature and/or probe hybridisation characteristics of the amplicon within the PCR reaction mixture. In the intercalating dye system the melting temperature of the amplicon can be estimated by measuring the level of fluorescence emitted by the dye as the temperature is increased from below to above the expected melting temperature. The methods that rely upon probe hybridisation to produce a fluorescent signal are generally less liable to produce false positive results than alternative methods such as the use of intercalating dyes to detect net synthesis of double stranded DNA (dsDNA) followed by melting analysis of the product. Hybridisation, ResonSense and hydrolysis probe systems give fluorescent signals that are only produced when the target sequence is amplified and are unlikely to give false positive results. An additional feature of the hybridisation, ResonSense and related methods is that it is also possible to measure the temperature at which the probes disassociate from their complementary sequences giving further verification of the specificity of the amplification reaction. An important feature of many of the probe systems is that they are compatible with multiplexing due to the availability of fluorophores with resolvable emission spectra. The chemistries available are discussed in detail in Chapter 3.

Real-Time PCR Instrumentation

Thermal cyclers with integrated fluorimeters and some arrangement for transferring excitation light from a source into the reaction vessel and then from the sample to a detector are required for real-time PCR. The heating blocks that are the mainstay of the standard PCR instrument market present several technical challenges in conversion to application in real-time machines. The main problem being that the light must be channelled through the lid of the block and the cap of the reaction vessel across an air gap and then into the sample. Emitted light must then take the return path. Although blocks are used by several real-time machines including the first commercial instrument (ABI 7700), the difficulties associated with them have led to the development of alternative designs. The LightCycler® (LC24) was the forerunner of machines that use air as the heating/cooling medium. Thermal transfer via air has the advantage of greater uniformity and rapidity than can be achieved on block-based cyclers, besides allowing shortening of the light path. Besides differing in the choice of heating medium real-time PCR machines also provide a range of options for the light source and detection of fluorescence. Current machines tend to allow the excitation and detection of multiple dyes so that internal standards and multiplex reactions are possible. There is also a tendency to build in a bias toward the use of either universal donor or universal recipient chemistry (see Chapter 3).

Since their introduction, the cost of real-time PCR instruments has fallen in tandem with continual improvement in their capability and accuracy. This has been the result of competition, the volume of sales and the introduction into the marketplace of improved designs dependent on new technology. These trends are unlikely to be reversed and will contribute to the growth in real-time PCR's popularity. Instrumentation for real-time PCR is described and discussed in more detail in Chapter 2.

Quantification

Unlike standard PCR, real-time PCR instruments measure the kinetics of product accumulation in each PCR reaction tube. Generally, no product is detected during the first few temperature cycles as the fluorescent signal is below the detection threshold of the instrument. However, most combinations of machine and fluorescence reporter are capable of detecting the accumulation of amplicons before the end of the exponential amplification phase. During this time the efficiency of PCR is often close to 100% giving a doubling of the quantity of product at each cycle. As product concentrations approach the nanogram per μl level the efficiency of amplification falls primarily because the amplicons re-associate during the annealing step. This leads to a phase during which the accumulation of product is approximately linear with a constant level of net synthesis at each cycle. Finally, a plateau is reached when net synthesis approximates zero. Quantification in real-time PCR is done by measuring the number of cycles required for the fluorescent signal to reach a threshold level or the second derivative maximum of the fluorescence versus cycle curve. This cycle number is proportional to the number of copies of template in the sample. Real-time quantification is discussed further in Chapters 6 and 7.

SNP Detection

The methods used to verify the identity of the amplicon(s) produced in real-time PCR are also often sufficiently powerful to detect small variations between sequences. Variations in sequence ranging from single nucleotide polymorphisms (SNPs) have been successfully identified in real-time PCR assays.

One common approach to the detection of sequence variation is to compare melting curves. In general, the effect of base substitutions on the melting kinetics of PCR products is too small to be detected reliably (if at all). However, one group (Wittwer *et al.*, 2003) has demonstrated that heteroduplexes of relatively long amplicons differing by a SNP can be distinguished from the homoduplexes on the basis of their melting curves. This was presented as the basis of a method for mutation screening. More commonly, the melting curves of short fluorescent probes are used to distinguish between amplicons, for example (Edwards *et al.*, 2001b; Whalley *et al.*, 2001). This method is sensitive to SNPs, which usually cause a shift in the melting peak of several degrees. A common alternative to the melting curve approach is to use hydrolysis (TaqMan) probes. The efficiency of the 5'-3' endonuclease reaction is greatly impaired when a well-designed probe mismatches its target sequence by even a single base. The detection of mutations by real-time PCR is discussed in Chapter 8. Although the melting curve and hydrolysis probe methods for mutation analysis are widely used they are only able to detect sequences that represent a large proportion of the population. The quantitative real-time ARMS assay described in Chapter 9 is designed to detect the emergence of significant sequence mutants within a background that remains mainly of the parent type.

Real-Time PCR Data Analysis

The software provided with real-time PCR instruments allows three principle types of data analysis. 1) Measurement of the cycle number at which any increase in the fluorescence within each reaction vessel reaches significance. 2) The data are used in conjunction with the results from external standards to estimate the original number of template copies. 3) Melting curves are transformed to provide plots of $-dF/dT$ against T (F = fluorescence and T=temperature) in which a peak (melting peak) occurs at the equilibrium temperature for each duplex. In general the different software is easy to use and allows rapid and reproducible data analysis.

Non-PCR Applications

Real-time PCR machines are also capable of use as real-time fluorimeters. For example, one simple application is estimation of the melting temperature (T_m) of an oligonucleotide. The oligonucleotide is mixed with its complementary sequence in the presence of a dye such as SYBR Green I, the temperature is increased and the level of fluorescence is measured to give a melting curve from which the T_m may be deduced.

Chapter 10 presents an alternative application using a real-time PCR instrument that relies on real-time fluorimetry. NASBA is a method for the isothermal amplification of RNA that produces quantities of antisense RNA copies. Molecular beacons complementary to the product are used to give a fluorescent signal.

The Growth in the Use of Real-Time PCR

In a relatively short time since their first introduction in the mid-1990s real-time PCR machines have become widely available to biologists. This has led to an explosion in the number of publications describing applications of the method. Indeed, a graph of number of papers against time resembles a real-time PCR plot (Figure 1). Most of the main applications that exploit real-time PCR previously relied on standard PCR and the main fields included diagnostic microbiology and human genetic analysis. However, the decreased hands-on time, increased reliability and improved quantitative accuracy of real-time PCR methods are contributing to a widening of their use into areas that were not previously dominated by PCR. For example, it has been

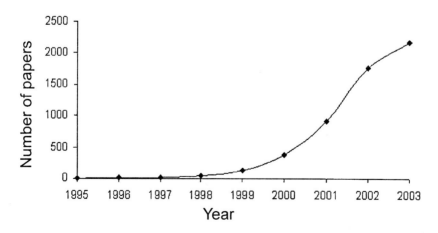

Figure 1. PubMed at NCBI (http://www.ncbi.nlm.nih.gov/) was searched by year for the term 'real-time PCR'. The result for 2003 was for an incomplete year.

exploited for gene expression analysis (Edwards and Saunders, 2001; Sabersheikh and Saunders, 2003).

Applications of Real-Time PCR

In recent years real-time PCR has found many biological applications. These can usually be classified as either quantitative or qualitative methods and according to whether the probes are used to distinguish between sequence variants or simply as reporters.

The simplest application of real-time PCR is for the detection of specific gene sequences within a complex mixture. Such assays are useful in the diagnosis of infectious disease where assays for DNA sequences specific for a wide range of pathogens have been developed. Several such applications are discussed in Chapters 11 and 12.

An important application of quantitative real-time PCR is the measurement of RNA transcript levels to assess gene expression. This application is making a critical contribution to our understanding of the interplay of host and microbe responses during infection. Gene expression studies can also help us to understand the functioning of normal tissues and to elucidate the pathogenesis of non-infectious diseases. The analysis of mRNA expression is discussed in Chapter 7. Other important applications of quantitative real-time PCR include the assessment of gene ratios in tumour tissues (Lehmann *et al.*, 2000; Kim *et al.*, 2002) and the measurement of pathogen numbers in clinical specimens (Brechtbuehl *et al.*, 2001; Eishi *et al.*, 2002).

Real-time PCR is frequently used for genotyping humans and human pathogens. In certain epidemiological studies the target sequences may be anonymous markers selected on the strength of their ability to discriminate between individuals or their linkage with particular phenotypes. More frequently target sequences are employed that are directly associated with a particular phenotype. Real-time PCR assays that use oligonucleotide probes for the detection of SNPs are now widely used. Examples include assays for bacterial identification (Edwards *et al.*, 2001a; Logan *et al.*, 2001), to detection of drug resistance (Edwards

9

et al., 2001b; Whalley *et al.*, 2001) and for diagnosis of human genetic disease (Costa *et al.*, 2003; Vrettou *et al.*, 2003).

References

Brechtbuehl, K., Whalley, S. A., Dusheiko, G. M., and Saunders, N. A. 2001. A rapid real-time quantitative polymerase chain reaction for hepatitis B virus. J. Virol. Methods 93: 105-113.

Costa, C., Pissard, S., Girodon, E., Huot, D., and Goossens, M. 2003. A one-step real-time PCR assay for rapid prenatal diagnosis of sickle cell disease and detection of maternal contamination. Mol. Diagn. 7: 45-48.

Edwards, K. J., Kaufmann, M. E., and Saunders, N. A. 2001a. Rapid and accurate identification of coagulase-negative staphylococci by real-time PCR. J. Clin. Microbiol. 39: 3047-3051.

Edwards, K. J., Metherell, L. A., Yates, M., and Saunders, N. A. 2001b. Detection of *rpoB* mutations in *Mycobacterium tuberculosis* by biprobe analysis. J. Clin. Microbiol. 39: 3350-3352.

Edwards, K. J., and Saunders, N. A. 2001. Real-time PCR used to measure stress-induced changes in the expression of the genes of the alginate pathway of *Pseudomonas aeruginosa*. J. Appl. Microbiol. 91: 29-37.

Eishi, Y., Suga, M., Ishige, I., Kobayashi, D., Yamada, T., Takemura, T., Takizawa, T., Koike, M., Kudoh, S., Costabel, U., Guzman, J., Rizzato, G., Gambacorta, M., du Bois, R., Nicholson, A. G., Sharma, O. P., and Ando, M. 2002. Quantitative analysis of mycobacterial and propionibacterial DNA in lymph nodes of Japanese and European patients with sarcoidosis. J. Clin. Microbiol. 40: 198-204.

Kim, Y. R., Choi, J. R., Song, K. S., Chong, W. H., and Lee, H. D. 2002. Evaluation of HER2/neu status by real-time quantitative PCR in breast cancer. Yonsei Med. J. 43: 335-340.

Kleppe, K., Ohtsuka, E., Kleppe, R., Molineux, I., and Khorana, H. G. 1971. Studies on polynucleotides. XCVI. Repair replications of short synthetic DNA's as catalyzed by DNA polymerases. J. Mol.

Biol. 56: 341-361.

Lehmann, U., Glockner, S., Kleeberger, W., von Wasielewski, H. F., and Kreipe, H. 2000. Detection of gene amplification in archival breast cancer specimens by laser-assisted microdissection and quantitative real-time polymerase chain reaction. Am. J. Pathol. 156: 1855-1864.

Logan, J. M., Edwards, K. J., Saunders, N. A., and Stanley, J. 2001. Rapid identification of *Campylobacter* spp. by melting peak analysis of biprobes in real-time PCR. J. Clin. Microbiol. 39: 2227-2232.

Sabersheikh, S., and Saunders, N. A. 2003. Quantification of virulence-associated gene transcripts in epidemic methicillin resistant *Staphylococcus aureus* by real-time PCR. Mol. Cell. Probes In press.

Saiki, R. K., Gelfand, D. H., Stoffel, S., Scharf, S. J., Higuchi, R., Horn, G. T., Mullis, K. B., and Erlich, H. A. 1988. Primer-directed enzymatic amplification of DNA with a thermostable DNA polymerase. Science 239: 487-491.

Saiki, R. K., Scharf, S., Faloona, F., Mullis, K. B., Horn, G. T., Erlich, H. A., and Arnheim, N. 1985. Enzymatic amplification of beta-globin genomic sequences and restriction site analysis for diagnosis of sickle cell anemia. Science 230: 1350-1354.

Vrettou, C., Traeger-Synodinos, J., Tzetis, M., Malamis, G., and Kanavakis, E. 2003. Rapid screening of multiple beta-globin gene mutations by real-time PCR on the LightCycler: application to carrier screening and prenatal diagnosis of thalassemia syndromes. Clin. Chem. 49: 769-776.

Whalley, S. A., Brown, D., Teo, C. G., Dusheiko, G. M., and Saunders, N. A. 2001. Monitoring the emergence of hepatitis B virus polymerase gene variants during lamivudine therapy using the LightCycler. J. Clin. Microbiol. 39: 1456-1459.

Wittwer, C. T., Reed, G. H., Gundry, C. N., Vandersteen, J. G., and Pryor, R. J. 2003. High-resolution genotyping by amplicon melting analysis using LCGreen. Clin. Chem. 49: 853-860.

2

An Overview of
Real-Time PCR Platforms

J. M. J. Logan and K. J. Edwards

Abstract

Real-time PCR continues to have a major impact across many disciplines of the biological sciences and this has been a driver to develop and improve existing instruments. From the first two commercial platforms introduced in the mid 1990s, there is now a choice in excess of a dozen instruments, which continues to increase. Advances include faster thermocycling times, higher throughput, flexibility, expanded optical systems, increased multiplexing and more user-friendly software. In this chapter the main features of each instrument are compared and factors important to weigh up when deciding on a platform are highlighted.

History of Real-time PCR

Today it is clear that few techniques have had such a powerful impact on biology than the development of the polymerase chain reaction (PCR). More recently the PCR has become even more sophisticated with the introduction of real-time PCR. Initial work by Higuchi and colleagues (Higuchi *et al.*, 1992) first demonstrated the simultaneous amplification and detection of specific DNA sequences in real-time by simply adding ethidium bromide (EtBr) to the PCR reaction so that the accumulation of PCR product could be visualised at each cycle. When EtBr is bound to double-stranded DNA and excited by UV light it fluoresces, therefore an increase in fluorescence in such a PCR indicates positive amplification. Soon afterwards they introduced the idea of real-time PCR product quantitation or 'kinetic PCR', by continuously measuring the increase in EtBr intensity during amplification with a charge-coupled device camera (Higuchi *et al.*, 1993). By creating amplification plots of fluorescence increase versus the cycle number, they demonstrated that the kinetics of EtBr fluorescence accumulation during thermocycling was directly related to the starting number of DNA copies. Fewer cycles are needed to produce a detectable signal, when a greater number of target molecules are present. Kinetic monitoring also provided a means whereby the efficiency of amplification under different conditions could be determined, providing for the first time insight into the fundamental PCR processes. Therefore, the principle underlying real-time PCR can simply be defined as the monitoring of fluorescent signal from one or more PCRs cycle-by-cycle to completion, where the amount of product produced during the exponential amplification phase can be used to determine the amount of starting material.

The approach described above was not ideal since EtBr binds non-specifically to DNA duplexes and non-specific amplification products, such as primer–dimers, can contribute to the fluorescent signal and result in quantification inaccuracies. Subsequent refinements, the most significant of which was the introduction of fluorogenic probes to monitor product accumulation, added a greater element of specificity to real-time PCR and provided greater quantitative precision and dynamic range than previous methods.

These significant advances to the basic PCR technique not surprisingly led to the development of a new generation of PCR platforms and reagents, which allowed simultaneous amplification and quantification of specific nucleic acid sequences cycle-by-cycle. Indeed a few years after Higuchi coined the term 'kinetic' or 'real-time PCR' the first commercial platforms were released on the market. The first was the Applied Biosystems ABI Prism 7700 Sequence Detection System, followed by the Idaho Technology LightCycler (now manufactured and sold by Roche Diagnostics) (Wittwer *et al.*, 1997) and its military field version, the Ruggedised Advanced Pathogen Identification Device (RAPID). Both of these platforms utilised fluorogenic chemistry and like any real-time PCR platform, they basically consist of a thermal cycler with an integrated optical detection system. New and improved models have now superseded these two instruments and several other manufacturers have introduced their own real-time PCR platforms. Choosing a suitable instrument is now a complex task. Real-time PCR offers many advantages that include:

- amplification and detection in an integrated system
- fluorescent dyes/probes allowing constant reaction monitoring
- rapid cycling times (20-40 mins for 35 cycles)
- high sample throughput (~200 –5000 samples/day)
- low contamination risk due to sealed reactions
- increased sensitivity (~ 3pg or 1 genome equivalent of DNA)
- detection across a broad dynamic range of $10 - 10^{10}$ copies
- reproducibility with a CV < 2.0 %
- allows for quantification of results
- software driven operation
- no more expensive than "in-house" PCR

Current disadvantages include:

- limited capacity for multiplexing using all chemistries
- development of protocols requires a high level of technical skill and/or support (research and development capacity and capital)
- high capital equipment costs
- analysis requires skill

This chapter is intended to provide an overview of the main features of real-time platforms, as well as highlighting aspects to consider when introducing real-time PCR to your laboratory. Information on the available fluorescent chemistries, their principles and methods is detailed in Chapter 3.

Real-Time PCR Platforms

A real-time PCR instrumentation platform consists of a thermal cycler, optics for both fluorescence excitation and emission collection, together with a computer and software for data acquisition and analysis. A wide range of systems are now available (see References) and these differ in their design and level of sophistication, providing users with several choices, which include: format, reaction vessels, emission and excitation wavelengths, throughput, level of control, chemistry, software, speed and applications. They all have in common the ability to measure the accumulation of PCR product during the exponential phase of the reaction using online fluorescence monitoring, whether specific or non-specific and hence provide accurate data on initial starting copy numbers. As amplification and detection are combined in a single step, the process can occur in a single closed reaction vessel, which eliminates any need for numerous post-PCR manual manipulations, as well as reducing the possibility of introducing contamination or variability. Additional technical advantages include both qualitative and quantitative PCR, mutation analysis, multiplexing and high throughput analysis. Although the fluorescence chemistries used in different platforms are similar, their mechanics and methodologies are wide ranging. A summary of the features of each platform is detailed in Table 1 and the reader is encouraged to refer to it throughout this chapter.

Real-Time PCR Thermocycling

The first component to consider in a PCR platform is the thermal engine. Successful thermal cycling is dependent on the accurate regulation of temperature in the sample vessels and the speed at which these target temperatures can be achieved. The majority of real-time platforms use advanced heating block technology based on the Peltier-effect, to actively transfer heat in and out of thin-walled plastic reaction vessels (*e.g.* ABI 7000, ABI 7900HT, DNA Engine Opticon2, Mx4000, Mx3000P, iCycler IQ). Peltier devices transfer heat from one side of a semiconductor to another. In general, blocks have significant mass and consequently a degree of thermal inertia.

Furthermore, the plastic insulating layer between the reaction vessel and the heater produces an additional thermal lag. As a consequence of this, the temperature transitions are relatively slow and blocks must be very carefully designed to minimise well-to-well variation. Other advances on the Peltier-based technology include its combination with Joule, resistive or convective technology to give improved temperature control and performance across the block. Three platforms employ alternative heat exchange technologies which permit more rapid thermal ramp rates than blocks, resulting in significantly increased thermocycling speeds. These include a stationary air-heated glass capillary format (LightCycler), a centrifugal air-heated plastic tube format (Rotor-Gene) and a high-thermal-conductivity ceramic heating plate plastic tube format (SmartCycler). For example, the time taken to equilibrate at 72°C using a Rotor-Gene is 0s compared to 15s with a standard 96-well block, resulting in run times that are on average 50% faster. Detailed information on the temperature specifications for each platform is shown in Table 2, which highlights that the LightCycler has the capacity to perform the fastest PCR and the Rotor-Gene has the smallest variation in temperature uniformity.

Real-Time PCR Optics

An integrated fluorimeter is required to detect and monitor the levels of fluorescence during the PCR process for real-time PCR and there is a range of options available for both the excitation light source and fluorescent emission detection. The light sources that cause fluorophore excitation can be classed as narrow- or broad-spectrum. If a broad-spectrum light source is employed (*e.g.* Mx4000, Mx3000P, iCycler IQ, ABI 7000) then filters can be used to provide light tuned to the excitation spectrum of a specific individual fluorophore. Such a system provides the user with a wider choice of available fluorophores, although it is best to select those with good separation of their emission spectra. A disadvantage of this optical system is that the light intensity passing through the filters can be limited and this could in theory limit the sensitivity of detection. There are currently two narrow-spectrum light sources used in real-time platforms, these can be light emitting diodes (LEDs) (*e.g.* LightCycler, DNA Engine

Table 1. Features of Real-Time PCR Platforms

Company Model	Applied Biosystems ABI Prism 7000	Applied Biosystems ABI Prism 7900 HT with automation accessory
Laser/Lamp	Tungsten halogen lamp	Argon ion laser
Detector	CCD camera	CCD camera
Thermocycling	Peltier element 9700 block	Peltier element 9700 block
Excitation spectrum	350-750 nm	488 nm
Filters/detection channels	4-position fixed filter wheel: FAM/ SYBR Green I, VIC/JOE, TAMRA and ROX	500-660 nm continuous wavelength detection
Format	96-well plates or 0.2 ml tubes	96- and 384-well plates (interchangeable blocks)
Time (40 cycles)	1.75 hours	1.75 hours
Reaction volume	25-50 μl	5-20 μl
Fluorescence chemistry	Hydrolysis probes, SYBR Green I, other chemistries possible but not supported	Hydrolysis probes, SYBR Green I, other chemistries possible but not supported
Multiplexing supported	2-plex hydrolysis probes	2-plex hydrolysis probes
Passive reference	ROX	ROX
Dimensions (H x W x D)	39 x 51 x 53 cm	64 x 125 x 84 cm
Weight	34 Kg	114 Kg
Other features	Primer/probe design software included	Primer/probe design software included
	PCR mastermix kits	PCR mastermix kits
	Parameter Specific Kits	Parameter specific kits
	Assay by design and assay on demand services	Assay by design and assay on demand services
		Robotic plate un/loading for up to 84 plates
		High throughput >5000 wells 8-hour day (30,000 endpoint)

Abbreviations:
LED, light emitting diode
CCD, charge-coupled device camera
PMT, photomultiplier tube
FAM, carboxyfluorescein
JOE, carboxy-4',5'-dichloro-2',7'-dimethoxyfluorescein
TAMRA, carboxytetramethylrhodamine
ROX, carboxy-X-rhodamine

Company Model	Roche LightCycler	Roche LightCycler Ver. 2
Laser/Lamp	LED	LED
Detector	3 photodetection diodes	6 photodetection diodes
Thermocycling	Heated air	Heated air
Excitation spectrum	470 +/- 10 nm	470 +/- 10 nm
Filters/detection channels	3 channels: 530, 640, 710 nm	6 channels: 530, 560, 610, 640, 670, 710 nm
Format	32 glass capillaries	32 glass capillaries
Time (40 cycles)	30 mins	30 mins
Reaction volume	20 μl	20 μl or 100 μl
Fluorescence chemistry	Hybridisation probes, hydrolysis probes, molecular beacons, SYBR Green I	Hybridisation probes, hydrolysis probes, molecular beacons, SYBR Green I
Multiplexing supported	2-plex hybridisation probes	4-plex hybridisation and 2-plex hydrolysis probes
Passive reference	Not required	Not required
Dimensions (H x W x D)	45 x 30 x 40 cm	
Weight	20 Kg	
Other features	Primer/probe design software available	Primer/probe design software included
	PCR mastermix kits	PCR mastermix kits
	Parameter specific kits	Parameter specific kits
	Relative quantification software available	Relative quantification software included
		Nucleic acid quantitation

Abbreviations:
LED, light emitting diode
CCD, charge-coupled device camera
PMT, photomultiplier tube
FAM, carboxyfluorescein
JOE, carboxy-4',5'-dichloro-2',7'-dimethoxyfluorescein
TAMRA, carboxytetramethylrhodamine
ROX, carboxy-X-rhodamine

Company Model	Stratagene Mx4000	Stratagene Mx3000P
Laser/Lamp	Quartz tungsten halogen lamp	Quartz tungsten halogen lamp
Detector	4 PMTs	1 scanning PMT
Thermocycling	Solid-state resistive/convective peltier hybrid block	Solid-state peltier-based block
Excitation spectrum	4 customisable filter wheels in the range 350-750 nm	4 customisable filter wheels in the range 350-750 nm
Filters/detection channels	4 customisable filter wheels in the range 350-830 nm	4 customisable filter wheels in range 350-700 nm
Format	96-well plates, 0.2 ml tubes, 8 x 0.2 ml strips	96-well plates, 0.2 ml tubes, 8 x 0.2 ml strips
Time (40 cycles)	1.5 hours	
Reaction volume	10-50 μl	25 μl
Fluorescence chemistry	Hydrolysis probes, molecular beacons, scorpions, amplifluor, SYBR Green I	All chemistries
Multiplexing supported	4-plex	4-plex
Passive reference	Optional ROX	Optional ROX
Dimensions (H x W x D)	76 x 46 x 51cm	33 x 46 x 43 cm
Weight	50 Kg	20 Kg
Other features	PCR mastermix kits	PCR mastermix kits
	Integrated computer	

Abbreviations:
LED, light emitting diode
CCD, charge-coupled device camera
PMT, photomultiplier tube
FAM, carboxyfluorescein
JOE, carboxy-4',5'-dichloro-2',7'-dimethoxyfluorescein
TAMRA, carboxytetramethylrhodamine
ROX, carboxy-X-rhodamine

Company Model	Cepheid SmartCycler	Corbett Rotor-Gene
Laser/Lamp	4 high intensity LEDs	4 high power LEDs
Detector	Silicon photodetectors	PMT
Thermocycling	Resistive heating of ceramic plates with forced-air cooling	Resistive heater with air cooling and centrifugation
Excitation spectrum	4 channels (450-495, 500-550, 565-590, 630-650 nm)	470, 530, 585, 625 nm
Filters/detection channels	4 channels (510-527, 565-590, 606-650, 670-750 nm)	510, 555, 610 580 hp, 610 hp, 660 hp nm (high pass)
Format	16 proprietary tubes	36 0.2 ml or 72 0.1 ml plastic tubes
Time (40 cycles)	40 mins	30 mins
Reaction volume	25 or 100 μl	10-100 μl
Fluorescence chemistry	Hydrolysis probes, molecular beacons, amplifluor and scorpion primers, SYBR Green I	Hydrolysis probes, molecular beacons, hybridisation probes, SYBR Green I
Multiplexing supported	4-plex	4-plex
Passive reference	Not required	Not required
Dimensions (H x W x D)	30 x 30 x 25 cm	31.5 x 38 x 48 cm
Weight	10 Kg	17 Kg
Other features	PCR mastermix kits	
	Random access (16 independent sites)	

Abbreviations:
LED, light emitting diode
CCD, charge-coupled device camera
PMT, photomultiplier tube
FAM, carboxyfluorescein
JOE, carboxy-4',5'-dichloro-2',7'-dimethoxyfluorescein
TAMRA, carboxytetramethylrhodamine
ROX, carboxy-X-rhodamine

Company Model	BioRad iCycler IQ	MJ Research DNA Engine Opticon2
Laser/Lamp	Tungsten halogen lamp	96 LEDs
Detector	CCD with intensifier technology	Dual PMTs
Thermocycling	Peltier-based	Peltier-based
Excitation spectrum	5 filter positions in range 400-700 nm	470-505 nm
Filters/detection channels	5 filter positions available, 2 provided	523-543, 540-700 nm
Format	96- or 384-well plate, 8 x 0.2 ml strip tubes	96-well plate, 8 x 0.2 ml strip tubes
Time (40 cycles)	2 hours	
Reaction volume	10-100 μl	10-50 μl
Fluorescence chemistry	Hydrolysis probes, molecular beacons, hybridisation probes, SYBR Green I	Hydrolysis probes, molecular beacons, hybridisation probes, SYBR Green I
Multiplexing supported	4-plex	2-plex
Passive reference	Fluorescein	Not required
Dimensions (H x W x D)	36 x 33 x 62 cm	60 x 34 x 47 cm
Weight	17.6 Kg	29 Kg
Other features	Primer/probe design software available	
	PCR mastermix kits	PCR mastermix kits
	Thermal gradient block	Thermal gradient block

Abbreviations:
LED, light emitting diode
CCD, charge-coupled device camera
PMT, photomultiplier tube
FAM, carboxyfluorescein
JOE, carboxy-4',5'-dichloro-2',7'-dimethoxyfluorescein
TAMRA, carboxytetramethylrhodamine
ROX, carboxy-X-rhodamine

Table 2. Temperature Specifications of Real-time PCR platforms			
Platform	Max. heating/ cooling rate (°C/sec)	Temperature accuracy (°C)	Temperature uniformity (°C)
ABI 7000/7900HT	1.5/1.5	+/- 0.25	+/- 0.5
iCycler IQ	3.3/2.0	+/- 0.3	+/- 0.4
LightCycler	20.0/20.0	+/- 0.4	+/- 0.2
Mx4000	2.2/2.2	+/- 0.25	+/- 0.25
Mx3000P	2.5/2.5	+/- 0.25	+/- 0.25
DNA Engine Opticon2	3.0/2.0	+/- 0.4	+/- 0.4
Rotor-Gene	2.5/2.5	+/- 0.5	+/- 0.01
SmartCycler	10.0/2.5	+/- 0.5	+/- 0.5

Opticon2, SmartCycler, Rotor-Gene) or laser (ABI 7900HT). The SmartCycler and Rotor-Gene each have four LEDs that excite at different wavelengths, providing a greater selection of fluorophores and giving these instruments capabilities similar to those of the broad-spectrum platforms above. The LightCycler, DNA Engine Opticon2 and ABI 7900HT have single light source excitation, which ultimately limits the choice of fluorophores.

In general, the detectors used in real-time platforms are set to measure narrow bands of the spectrum, although filter sets that can be customised by the user are available for the iCycler IQ, Mx4000 and Mx3000P. The number of detection channels that can be effective is dependent on the available range of excitation wavelengths. For example, if a single narrow range excitation source is available, one approach is to use fluorophores that are all excited to some extent in the same range and then to rely on software correction to deconvolute the light emitted from a given area of the spectrum, as was employed successfully with the now discontinued ABI 7700. Another approach is demonstrated with the LightCycler, where a narrow-spectrum light source excites the fluorophores SYBR or fluorescein and emitted light is collected via three discrete optical detectors. Two of these detect long wavelength light emissions from fluorophores which are only minimally excited by the blue LED light source, but which are instead excited using FRET technology (see Chapter 3).

Real-Time PCR Chemistries

As the technology has advanced rapidly, second and third generation real-time PCR platforms have been developed with improvements seen in multiplexing and increased throughput capabilities. The optical characteristics of a given platform clearly have an impact on the ability to multiplex and also determine which probe systems are compatible. In addition, the analysis software may also predetermine the appropriate chemistries. The ability to mulitiplex the available fluorophores is fully discussed in Chapter 3. However, it is important to point out that the platform and choice of fluorescent chemistry are strongly linked. Indeed some platforms are biased towards a particular probe system and whilst the optics permit different probe chemistries to be excited and detected, often the analysis software does not and the user is required to export the data to a spreadsheet program for detailed user analysis. For example, the Applied Biosystems platforms do not officially support any chemistries other than hydrolysis probes. Therefore, the reporting chemistry required for an application should be strongly considered before a choice of platform is made.

Additional Platform Features

Several of the platforms employ a standard 96-well block format or interchangeable 384-well block and offer a medium to high throughput. An advantage of employing a block format is that standard PCR plates and tubes can be used and these tend to be cheaper than instrument-specific plastics (SmartCycler, Rotor-Gene) or glass capillaries (LightCycler). Also, the LightCycler and SmartCycler require centrifugation to move sample into the reaction vessel and these alternative designs may not be suitable for all applications. For the highest throughput, the ABI 7900HT combines a 384-well plate format with an automation accessory, which allows for up to 84 plates to be loaded and unloaded, providing a throughput of >5000 wells per 8-hour day or 30,000 wells for end-point analysis only. These high throughput instruments are ideal for dedicated laboratories where large batches of samples are run, with few different cycling parameters. An additional specification of a few platforms is the ability to perform gradient thermocycling, which can be very useful at the assay optimisation stage.

Some platforms offer a low to medium throughput but are more flexible. Although the format may allow only 32 or even 16 samples, thermocycling times are faster and multiple runs can be performed thereby increasing the potential throughput. Performing multiple runs rapidly may be an advantage when several applications are employed which require different cycling parameters. The SmartCycler offers a new concept to real-time PCR of random access, which means independent programming of cycling parameters for 16 different assays. Each reaction vessel has its own element so that runs in available slots may be started any time, whilst other reaction sites are already in use.

Ideally, the analysis software supplied with the platform should be as user-friendly as possible but it is also important to check that the software can fully analyse results of the chosen probe chemistry. As already mentioned, some platforms and analysis software suites are biased towards certain chemistries. Some real-time instruments also have specific primer and probe design software that is either supplied with the hardware or available at extra cost. Such software can help simplify and speed up the assay design process and is optimised for that system and reagents. The LightCycler also has specific relative quantification software that is designed to determine the exact relative nucleic acid concentration normalized to a calibrator sample. This software speeds up and greatly simplifies this method of quantification.

The majority of instrument manufacturers supply optimised real-time PCR mastermixes, these reagents benefit from being quality controlled, are easy to use, and usually offer reproducible and reliable results. However, in certain laboratories cost can be prohibitive, although with other companies (*e.g.* Epicentre, Eurogentec, Invitrogen, TaKaRa, Qiagen) now supplying real-time reagents, competition in this market should lead to reduced costs. Target specific kits for a range of applications are available from some manufacturers and other companies (*e.g.* Applied Biosystems, Artus Biotech, AME Bioscience, Idaho Technologies, Minerva Biolabs, Roche, TaKaRa). Additionally, Applied Biosystems offer Assays-on-Demand and Assays-by-Design services for SNP genotyping by real-time PCR.

In the current world climate there has been a drive in the real-time PCR market towards portability, sensitivity and rapid response capabilities. Currently the RAPID and a portable version of the SmartCycler are available as field deployable real-time PCR machines, together with a range of freeze-dried PCR reagents and specific detection kits. They also provide simple 'push button' software which permits use by personnel with minimal training. A range of other handheld PCR biodetection devices have also been developed with the threat of bioterrorism in mind, (see also Chapter 11) such as: the Miniature Analytical Thermal Cycling Instrument (Northrup *et al.* 1998); the Advanced Nucleic Acid Analyser (Belgrader *et al.,* 1998); the Handheld Advanced Nucleic Acid Analyser (Lawrence Livermore National Laboratories) and its commercial counterpart the Smiths Detection Bio-Seeq; and the Idaho Technology RAZOR instrument. Clearly, these instruments are designed with specific application requirements and their evaluation and implementation will require careful consideration.

Finally, a new introduction to the marketplace is the BioGene InSyte real-time platform, which uses a novel electrically conducting polymer technology (ECP). The ECP is formed into a 96-well moulded thermal plate, which allows ultra rapid thermocycling due to the low thermal mass combined with a high surface:area ratio. As there is temperature measurement of each tube, well-to-well uniformity issues have been eliminated. This platform also has significant flexibility as each tube is the heating element and can be thermally addressed for individual thermal cycling. The optics consists of a 473nm laser and a 32-channel spectrometer (520-720nm) with no filters, where each channel is a photomultiplier tube (PMT). This instrument should allow the majority of chemistries to be utilised. Since this is a relatively new instrument experience and applications with this platform are currently unavailable.

Final Considerations

In weighing up the pros and cons of the different platforms for your laboratory, factors to consider include: supported chemistries; multiplex capability for that chemistry; throughput; flexibility; format; easy-to-use and robust software package; reproducibility; speed; size; technical support; customer support and not least the cost, not only of the initial equipment outlay and servicing but also the associated cost of consumables and reagents. It is also useful to 'try before you buy', most companies tend to provide a loan machine and if possible try to test a few of these once you have narrowed down your choice. User experiences should not be overlooked and there are now a number of useful websites and news groups where you can address you questions and queries (see References).

Clearly real-time PCR has undergone significant developments over the last ten years, which has resulted in a wide range of different platforms becoming available. The drive to improve current technology is likely to continue for the foreseeable future as real-time PCR finds it way into even more laboratories and specialist niches such as bioweapon sensing. As competition intensifies, users should benefit from more sophisticated high-throughput platforms and portable devices, with increasing levels of automation and cost-effectiveness, although making a choice of platform will continue to be difficult.

References

Papers

Belgrader, P., Benett, W., Hadley, D., Long, G., Mariella, R. Jr., Milanovich, F., Nasarabadi, S., Nelson, W., Richards, J., and Stratton, P. 1998. Rapid pathogen detection using a microchip PCR array instrument. Clin. Chem. 44: 2191-4.

Higuchi, R., Dollinger, G., Walsh, P.S., and Griffith, R. Simultaneous amplification and detection of specific DNA sequences. 1992. Biotechnology (N Y). 10: 413-417.

Higuchi, R., Fockler, C., Dollinger, G., and Watson, R. 1993. Kinetic PCR analysis: real-time monitoring of DNA amplification reactions. Biotechnology (N Y). 11: 1026-1030.

Northrup, M. A., Benett, B., Hadley, D., Landre, P., Lehew, S., Richards, J., and Stratton, P. A. 1998. A miniature analytical instrument for nucleic acid acids based on micromachined silicon reaction chambers. Anal. Chem. 70: 918-22.

Wittwer, C. T., Ririe, K. M., Andrew, R. V., David, D. A., Gundry, R. A. and Balis, U. J. 1997. The LightCycler: a microvolume multisample fuorimeter with rapid temperature control. Biotechniques. 22: 176-181.

Manufacturers' Websites

Applied Biosystems **www.appliedbiosystems.com**
AME Bioscience **www.amebioscience.com**
Artus Biotech **www.artus-biotech2.com**
BioGene **www.biogene.co.uk**
Bio-Rad Laboratories **www.bio-rad.co.uk**
Cepheid **www.smartcycler.com**
Corbett Research Ltd **www.corbettresearch.com**
Epicentre **www.epicentre.com**
Eurogentec **www.eurogentec.com**
Idaho Technology **www.idahotech.com**
Invitrogen **www.invitrogen.com**
Minerva Biolabs **www.minerva-biolabs.com**
MJ Research Inc. **www.mjr.com**
Qiagen **www.qiagen.com**
Roche Diagnostics Ltd **www.roche-applied-science.com**
Smiths Detection **www.smithsdetection.com**
Stratagene **www.stratagene.com**
TaKaRa **www.takarabioeurope.com**

Web Resources

http://130.241.73.140/archive
www.lightcycler-online.com
www.llnl.gov
www.meltcalc.de
www.tataa.com
www.wzw.tum.de/gene-quantification

News Groups

http://groups.yahoo.com/group/lightcycler/
http://groups.yahoo.com/group/qpcrlistserver/
http://groups.yahoo.com/group/realtimepcr/
http://groups.yahoo.com/group/taqman/

3

Homogeneous Fluorescent Chemistries for Real-Time PCR

M. A. Lee, D. J. Squirrell, D. L. Leslie, and T. Brown

Abstract

The development of fluorescent methods for a closed tube polymerase chain reaction has greatly simplified the process of quantification. Current approaches use fluorescent probes that interact with the amplification products during the PCR to allow kinetic measurements of product accumulation. These probe methods include generic approaches to DNA quantification such as fluorescent DNA binding dyes. There are also a number of strand-specific probes that use the phenomenon of Fluorescent Resonance Energy Transfer (FRET). In this chapter we describe these methods in detail, outline the principles of each process, and describe published examples. This text has been written to provide an impartial overview of the utility of different assays and to show how they may be used on various commercially available thermal cyclers.

Introduction

A fluorescent real-time polymerase chain reaction (PCR) (Saiki *et al.*, 1985; Mullis *et al.*, 1987) can provide both qualitative and quantitative analysis for various applications. Real-time PCR differs from earlier methods of analysis in that additional components are required to carry out the process. These components include an optical system integrated into the thermal cycler, and a probe that reports amplification during the course of the PCR process. The real-time PCR thermal cycler is discussed in chapter 2 in detail. In this chapter we discuss the probe technologies or "reporting chemistries."

In order to give the reader a complete understanding of current technology, the principles of real-time analysis will initially be presented outside the context of any specific commercial platform. The number of these is increasing and it is certain that their capabilities on offer will undergo continued improvement and allow new fluorescent approaches to be realised. At this point it is important to highlight to new users the relationship between the choice of probe system and the instrument. These are inextricably linked: the optical specification and the analysis tools on any given platform greatly influence the applicability and utility of different probe systems. Whilst the main factors for the choice of instrument are often driven by throughput requirements and the initial purchase cost, careful consideration of the reporting chemistry for the required application should be made before purchase since the operation of one chemistry or another may be greatly compromised on some platforms. Equally important is the technical support provided from suppliers. This may be limited for non-supported chemistries and so called "open platforms" from manufacturers that do not support any one chemistry. These suppliers may not be able to provide technical advice for the chemistry of choice for the required application. In this chapter we break down probe technology into a number of components to enable the reader to understand how the different assay systems may be used on current and future fluorimetric thermal cyclers. For new users this will facilitate the implementation of various assays on commercial platforms.

We first present a background to the limitations of PCR without real-time detection and introduce the two main classes of real-time chemistries. A second section on the basics of Fluorescent Resonance Energy Transfer (FRET) will allow the reader to understand how dye technologies may be applied in strand-specific applications. An outline of the function and utility of real-time PCR probe methods will provide the reader with an understanding of how such technology may be best applied. Finally, experimental considerations and specific examples will be provided although the reader should refer to the relevant chapters in this text for detailed information.

Background to Real-Time PCR

Prior to the introduction of the first commercial real-time PCR instruments in 1996 the utility of the PCR was limited. It could only be easily applied as a qualitative method. Analysis of amplification products was carried out at the end of thermal cycling using techniques such as gel electrophoresis or PCR ELISA. The dynamic range for quantitative PCR using these approaches was limited because the accumulation of specific product plateaus when the reaction is allowed to progress to completion. At high cycle numbers the amount of product is often unrelated of the initial amount of target nucleic acid. Factors that contribute to the plateau effect include, amongst others, substrate depletion (nucleotides, amplimers, etc), specific inhibition (competitive binding of products-to-products rather than amplimer-to-product) and non-specific product inhibition (amplimer artefact accumulation and mis-priming), and the accumulation of pyrophosphate. Quantitative analysis, with a limited dynamic range, could only be achieved by stopping thermal cycling before this plateau was achieved. Competitive PCR (utilising molecular mimics that compete for amplimers) could be used to achieve a wider dynamic range. However, this approach is less accurate and more time consuming than the real-time methods described here.

The fluorescent real-time approach requires a thermal cycler that can interrogate the sample throughout the course of amplification. The signal collected during the exponential stage of amplification is of most use for the determination of the number of initial target nucleic acid molecules. At this stage the reaction efficiency is so high that the number of amplicon molecules effectively doubles every cycle. Amplification is observed as a two-fold increase in signal each cycle as it rises above that of the background noise. The amount of signal noise in a real-time assay is a function of the type of chemistry utilised, the optical performance of the instrument as well as other experimental considerations such as a volume effects and the degree of mixing of reaction cocktails.

There are two main classes of real-time fluorescent chemistries: "generic" and "strand-specific" methods. Generic methods use probes that bind non-specifically to DNA and include the intercalators and other DNA binding dyes. Strand-specific methods use nucleic acid probes that target the amplicon (product) between the amplimer binding regions. It should be noted that most of the probes described utilise DNA for the probe. The use of modified bases was proposed in 1991 (Holland *et al.*, 1991). Methods for peptide nucleic acid (PNA) probes utilising FRET have been reported (Ortiz *et al.*, 1998) and described for application in real-time PCR (Svanik *et al.*, 2000a), and for those probes that use hybridisation alone (rather than hydrolysis) it should be possible to utilise PNA and other analogs such as locked nucleic acid (LNA) (Koshkin *et al.*, 1998) in the chemistries. The strand-specific methods have the advantage that amplification of reaction artefacts such as amplimer-dimers do not contribute to the observed signal. Generally they provide higher specificity and better signal:noise ratios than the generic methods that are described next. The type and mode of action of different probe systems is illustrated in Table 1.

Generic Detection Using DNA Binding Probes and Melting Point Analysis

DNA binding dyes have been used extensively in molecular biology research for the direct analysis and quantification of nucleic acids. These dyes bind to dsDNA with enhanced fluorescence. Ethidium bromide, which is used conventionally for staining agarose gels, has been used in real-time PCR (Higuchi *et al.*, 1992; Higuchi *et al.*, 1993).

Table 1. Summary of fluorogenic reporting chemistries for real-time PCR. The schematics show the probe structure and mode of action.

Method	Principle	Self probing equivalent	Probe Structure	Nucleic Acid interaction
DNA binding agents	Enhancement of Fluorescence	n/a		
5' Nuclease Assay	Hydrolysis of Probe	Intrataq		
Dual Hybridisation probes	Hybridisation of linear probes	Embodiment of Scorpions		
ResonSense	Hybridisation of linear probes	Angler		
Hybeacons	Hybridisation of linear probes	Not reported		
Eclipse	Hybridisation of linear probes with conformational change	Not reported		
Light-up	Hybridisation with intercalator	Not reported		
Molecular Beacon	Hybridisation and conformational change	Scorpions		
Ying-Yang Probes	Competative binding of linear probes	Duplex Scorpions		

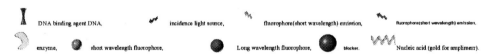

I DNA binding agent DNA.   incidence light source,   fluorophore(short wavelength) emission,   fluorophore(short wavelength) emission,   enzyme,   short wavelength fluorophore,   Long wavelength fluorophore,   blocker.   Nucleic acid (gold for amplimerr).

Others, such as those from Molecular Probes (YoPro® and YoYo®) (Ogura *et al.*, 1994), have also been reported. The fluorescent signal is usually monitored towards the end of the extension step in a 3 step PCR.

Of these probes by far the most reported are the minor-groove binding dyes SYBR® Green I (Becker *et al.*, 1996) and SYBR® Gold. These

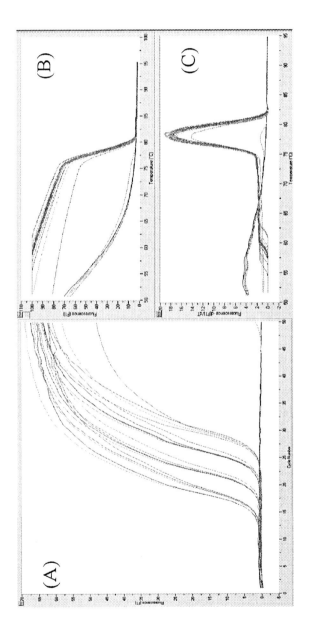

Figure 1. Amplification using a generic binding dye (minor-groove binding dye SYBR® Gold) on the LightCycler®. A: The amplification plot shows 50 cycles (to completion) of amplification of four dilutions (4 replicates) of a 10-fold dilution series and four no-template controls, one of which produces a low amplification signal. The reaction utilised anti-Taq antibody hot start and UNG carry-over protection and the data analysis utilises the background subtraction algorithm. B: Melting point analysis of the amplified products. C: The first negative differential of the fluorescence, with respect to temperature, plotted against temperature. From this plot it can be observed that the signal in the positive no template control has the same melting point as the specific amplification in positive samples, and therefore amplification was most likely a result of cross-contamination.

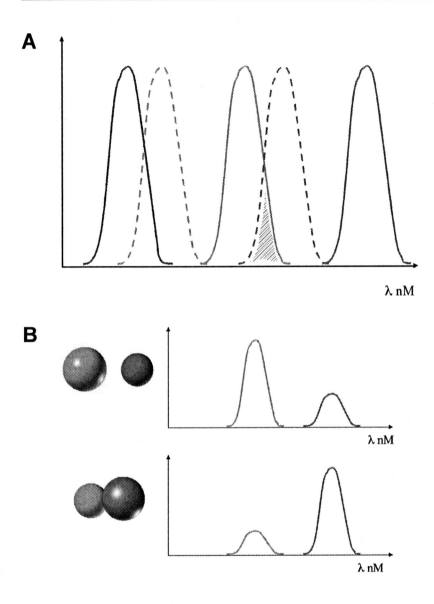

Figure 2. Fluorescence Energy Resonance Transfer. A: The emissions of a light source (blue line) falls within the excitation spectrum (green line dashed) of a fluorophore and cause this dye to fluoresce. The emissions (green solid) of this fluorophore fall within the excitation spectrum (red dashed line) of a second fluorophore that emits fluorescence at longer wavelengths (red solid). When the second fluorophore is close to the first on the same or neighbouring molecule, the energy that would be emitted by the first fluorophore (B top) can be transferred to the second through a number of energy transfer mechanisms such that the second fluorophore emits this energy at its own emission wavelengths (B bottom).

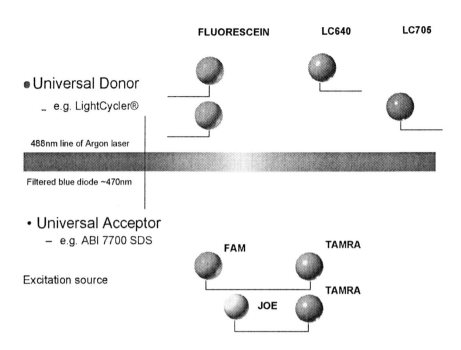

Figure 3. Schematic showing the two principal methods for multiplexing using FRET. Top, the Universal Donor on the LightCycler®, and bottom, the Universal Acceptor on the ABI 7700.

dyes typically exhibit 20-100 fold fluorescence enhancement on binding to dsDNA and are commonly used because their emission maxima closely match that of fluorescein and the optics in most commercial real-time instruments are set to detect in this (circa 520nm) wavelength range. The SYBR® dyes are popular for real-time detection because they are readily available from PCR reagent suppliers and their use requires little additional experimental design.

The optimum concentration for these dyes for a number of instrument platforms is published in the open literature. For SYBR® dyes this is typically 1:30,000 to 1:100,000 dilution of the reference solution supplied by the manufacturer, although the concentration is dependent on the tube format (composite glass/native polypropylene) and the optical efficiency of the instrument. Ready-to-go cocktails are available from suppliers with the dye already in the reagent mix.

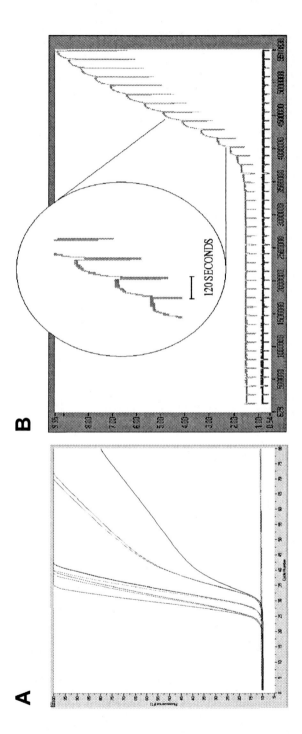

Figure 4. Amplification using the 5' nuclease assay on the LightCycler®. A: The amplification plot shows 80 cycles (in excess of completion) of thermal cycling of three dilutions (3 replicates) of a 10-fold dilution series and three no-template controls on the Roche LightCycler®. The reaction utilised anti-Taq antibody hot start and UNG carry-over protection. The reaction used the primers and probes from the Applied Biosystems Human β-Actin kit which utilises a FAM-TAMRA internally quenched TaqMan® probe. The amplification plot shows raw data (no analysis/background subtraction) and illustrates the excellent signal: noise ratio achievable with this chemistry. B: Shows the same reaction amplified on the Idaho LightCycler® carried out by continuous monitoring of a positive and negative sample throughout all stages of amplification. The data shows that ~120 seconds are required each cycle to take the hydrolysis of bound probe to completion (plateau) each cycle.

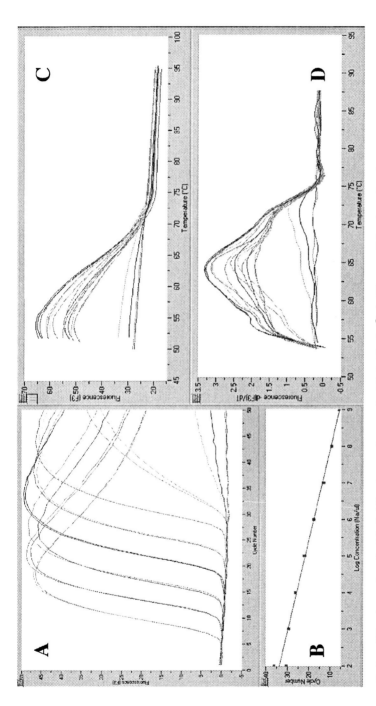

Figure 5. Amplification using dual hybridisation probes on the LightCycler®. A: The amplification plot shows 50 cycles of thermal cycling of replicates of 8 dilutions of a 10-fold dilution series of a reference for the target, and no-template controls on the Roche LightCycler®. The amplification shows the data analysed using the background subtraction algorithm. The reaction utilised anti-*Taq* antibody hot-start and UNG carry-over protection. The reaction primers and probes for the Human β-globin gene using a 3' labelled fluorescein donor probe and a 5' Cy5.5 acceptor probe. The acceptor probe was blocked to prevent 3' extension using an octendiol moiety. A 5'-3' exonuclease *Taq* polymerase was used to illustrate that the "hook" effect that is observed in linear hybridisation probes as a result of probe displacement by product in later cycles. B: The derived standard curve from the amplification plot. C: The melt curve showing probe disassociation from product. D: The first negative differential of fluorescence with respect to temperature showing the peak at which the maximum rate of probe dissociation occurs.

40

Table 2. Some of the common fluorophore combinations that can be used effectively in energy transfer pairs.

Fluorophore	Excitation Maxima (nm)	Emmission Maxima (nm)	As a Donor when used with	As an Acceptor when used with
Cascade Blue	396	410	DABCYL	
Cy3	552	570	DABCYL, BHQ-2, ElleQuencher, Eclipse	Fluorescein, SYBR®Gold, SYBR®Green-1
Cy3.5	581	596	DABCYL, ElleQuencher	Fluorescein, SYBR®Gold, SYBR®Green-1
Cy5	643	667	DABCYL, BHQ-2, ElleQuencher, Eclipse	Fluorescein, SYBR®Gold, SYBR®Green-1
Cy5.5	675	694	DABCYL, BHQ-2, ElleQuencher	Fluorescein, SYBR®Gold, SYBR®Green-1
EDANS(5-((2-aminoethyl)amino)naphthalene-1-sulfonic acid)	335	493	DABCYL	n/a
FAM (carboxyfluorescein)	494	518	TAMRA, ROX, Cy3, Cy3.5, Cy5, Cy5.5, DABCYL, BHQ-1/2, Eclipse	n/a
Fluorescein	492	520	TAMRA, ROX, Cy3, Cy3.5, Cy5, Cy5.5, DABCYL, BHQ-1/2, Eclipse	n/a
HEX (carboxy-2',4,4',5',7,7'- hexachlorofluorescein)	535	556	TAMRA, DABCL, BHQ-1/2, Eclipse	n/a
JOE (carboxy-4', 5'-dichloro-2',7'-dimethoxyfluorescein)	520	548	TAMRA, DABCL, BHQ-1, ElleQuencher	n/a
LC640	625	640	n/a	Fluorescein
LC705	685	705	n/a	Fluorescein
ROX (carboxy-X-rhodamine)	585	605	DABCYL, BHQ-2, ElleQuencher	Fluorescein
SYBR® Gold	495	537	Cy5, Cy5.5, LC 640, LC705	n/a
SYBR® Green-1	494	521	Cy5, Cy5.5, LC 640, LC705	n/a
TAMRA (carboxytetramethylrhodamine)	565	605	DABCYL, BHQ-2, ElleQuencher	Fluorescein
TET (carboxy-2',4,7,7'- tetrachlorofluorescein)	521	544	TAMRA, DABCL, BHQ-1/2, Eclipse	n/a
Texas Red	583	603	BHQ-2	n/a
VIC	538	554	TAMRA	n/a
Yakima Yellow	526	549	Eclipse	n/a

NOTE: DABCYL(((4-(dimethylamino)phenyl)azo)benzoic acid), Black Hole Quencher 1 &2 (BHQ), ElleQuencher™ and Eclipse™ are non-fluorescent "dark" quenching moeities.

41

Binding dye chemistries are excellent for assay optimisation. The signal is proportional to the total nucleic acid concentration and therefore directly related to the PCR process. In an optimised assay, they may thus be used to determine accurately the reaction efficiency. It should be noted that this is not always the case when using labelled nucleic acid probes, where the signal generated is related to both the assay and the "probe efficiency". In some assay systems the probe efficiency can have a major effect on the signal obtained and this is discussed later for each assay type. In both strand-specific and generic real-time PCR, the signal obtained is additionally dependent on the efficiency of the reverse transcriptase step.

Melting point analysis may be carried out on most commercial real-time instruments, usually at the end of the PCR amplification (Figure 1). Temperature dependent fluorescence measurements are made whilst slowly increasing the temperature of the reaction products from around 50°C to around 95°C. As DNA duplexes melt apart the fluorescence decreases as the bound dye is released. Most instruments provide an analysis of this data by plotting the first negative differential of the fluorescence signal with respect to temperature, against temperature. This plot appears as one or more peaks representing the point(s) at which the maximum rate(s) of change in fluorescence occur corresponding to a particular dsDNA product. Specific reaction products generally melt at a higher temperature than artefacts such as amplimer dimers and mis-primed products. Melting temperature is a function of GC content and to a lesser extent product length, which allows the converse to be derived in some reactions.

The melting peak method is analogous to that of agarose gel electrophoresis in that it does not unequivocally determine the presence of the correct sequence. Products with similar molecular masses or melting peaks may not be resolved by either technique. Gel electrophoresis has a higher resolution than melting point analysis and the quality of the melting point data collected varies greatly between hardware platforms. Those instruments where the temperature is accurately controlled and recorded, and where the correlation between temperature and fluorescence is good, will produce the highest resolution melting point data.

The major limitation of chemistries based on binding dyes is that the dyes bind to total nucleic acid, and in a PCR of sub-optimal efficiency a significant amount of signal will be derived from reaction artefacts generated as a result of ectopic amplimer hybridisation and enzymase extension. This leads to a dramatic loss in the dynamic range of quantification of samples and standards with low (<1000) initial target copies. However, a melting point analysis allows non-specific artefacts to be discriminated, to some degree, from specific amplicons (Ririe *et al.*, 1997).

The melting point analysis adds significantly to the utility of fluorescent DNA binding dyes since the data can be used to determine a number of important variables. Amplimer complements can be synthesised and used in melting point experiments to empirically determine annealing temperatures. The melting temperature of the product when determined may be used to improve the reaction in two respects on some instruments. Firstly, sequentially decreasing the denaturing step to a few degrees above the product melting temperature ensures that high molecular mass sample DNA is not denatured in later cycles which reduces the risk of ectopic priming events in later cycles. Secondly, this lowering of the denaturing temperature allows a shortening of the time taken for thermal cycling. If the melting temperature of reaction artefacts is significantly lower than that of the amplicon then it is possible to monitor fluorescence at a temperature just below the melting point of the amplicon, but higher than that of artefacts (Morrison *et al.*, 1998). This serves to reduce the artefact contribution to the specific amplification signal and may improve the dynamic range of quantification for a given assay.

An important consideration when using such dyes is the incorporation of a "hot-start" to reduce the formation of artefacts and so improve the dynamic range of quantification (Morrison *et al.*, 1998; Lee *et al.*, 1999). It should be noted that most of these reagents are both toxic and mutagenic to varying degrees, due to their ability to bind DNA. Storage conditions for these reagents are critical since they may be photo-bleached by both ambient and artificial light. The use of light-tight containers such as darkened microfuge tubes will reduce this, but most are labile in long-term storage.

When optimised a SYBR® Green I or SYBR® Gold assay can be used for most applications (such as expression studies). The ability to multiplex these assays is limited because the binding, which generates the fluorescence, is non-specific. However, it is possible to multiplex by melting point providing that the melting peak of each species can be resolved. Whilst allelic variation and mutation detection can be carried out, for example, using the amplification refractory mutation system (ARMS) approach (Newton *et al.*, 1989), this requires that assays for both variants be carried out in separate tubes. Because of these limitations assays that require high confidence results are best addressed using the strand-specific approaches that are discussed next.

Strand-Specific Detection Using Labelled Nucleic Acid Probes

The strand-specific methods employ fluorophore-coupled nucleic acids to interact with reaction products, probing accumulating PCR products for the presence of the target sequence. Such PCRs exhibit greater specificity than the respective assay without the probe component. However, the probe is targeted for only a portion of the amplicon sequence and therefore the signal for the product only reports the presence of the complementary sequences rather than the amplicon itself. The basis of detection of the product can be classed by the mode of action of the probe.

The efficiency of the probe is important if it is to produce a signal that correlates closely with the concentration of amplicon in the exponential stage of amplification and thus is used to determine the number of initial target copies. With an inefficient probe the signal may not increase exponentially and not correlate with initial target copy numbers. Inefficient probes will produce an inaccurate value for the PCR efficiency. The signal generated by a probe is dependent on, for example, sufficient hybridisation and/or cleavage as is later described. The target sequence is the main factor affecting probe efficiency because of different sequence dependent factors for a given probe type. Therefore the benefits of selecting a probe type should be carefully considered both when designing an experiment and purchasing an instrument since the preferred assay system may not easily be applied on every instrument.

The optimum composition of the core chemistry may vary with different platforms and this is discussed in chapter 4. The optical specifications of platforms differ such that they will be optimised for specific fluorphores. The concentration of probes may also vary between platforms but typically probes are used at a final concentration of around 0.1-1 μM.

In this text probe types are grouped for discussion by mode of action. The evolution of real-time methods is not presented chronologically, although reference is made to how and when they were implemented. Most of these methods use the phenomenon of FRET. This phenomenon will be discussed first and then the application and experimental considerations of the various methods will be outlined.

Fluorescent Resonance Energy Transfer Probes and the Passive Reference

FRET involves the non-radiative (no photon emitted) transfer of energy between fluorophores and has been used for a number of biological assays. There are excellent reviews of principles and applications (Cardullo, *et al* 1988; Selvin 1995; Clegg, 1992), so there is no need to include an in depth description of the process in this text. Here we present an overview and the two main approaches to detection and multiplexing using FRET as used in real-time PCR experiments. Reference will be made to specific instruments to illustrate the process.

For FRET to occur there must be at least two fluorophores in close proximity. One fluorophore called the "donor," is excited by an incident light source and emits light at a longer wavelength that falls within the excitation spectrum of the second fluorophore, called the "acceptor." When the donor and acceptor are on the same or neighbouring molecule then the emission energy of the excited donor may be transferred

45

through a number of distance and orientation dependent effects including dipole-dipole interactions between the donor and acceptor. Variations in the intensity of emissions of fluorophores on nucleic acids resulting from spatial changes during the course of amplification provide the basis for detecting amplicon accumulation (Figure 2). The choice of dye pairs is important in achieving a good signal:noise ratio. Since the Förster distance (the distance between the dye pair at which the efficiency of energy transfer is 50%) varies for different dye combinations.

Analysis utilising FRET involves monitoring either a rise or drop in fluorescence of either fluorophore depending on the assay type and instrument. On some platforms it is possible to present this data as a change in the quotient of the emissions of the two dyes. For example, if the basis of detection is the donor fluorophore being liberated spatially from the acceptor then one could express the data as donor/acceptor. If the converse is true the basis of detection is the donor being brought in close proximity to the acceptor one could express the data as acceptor/donor. This simple type of analysis is useful in a non-multiplexed real-time PCR since subtle changes in sample volume due to pipetting errors and/or changes in sample volume due to evaporation can be normalised out.

Of importance to the signal:noise ratio of assay systems utilising FRET is the ratio of dyes in a FRET pair. In particular this is important where two dyes may be used on the same probe (dual labelled) as described later. These probes are more complicated in terms of manufacture since they require incorporation of both dyes into the one probe molecule and are therefore more expensive to synthesise. Oligonucleotide probes are usually made by one of two common methods. The first dye, if this is at the 3'-end of the probe, is part of the support matrix on which the probe is synthesised by sequential addition of each base in the sequence. The second dye is either directly incorporated by the use of fluorophore labelled monomer, or by the inclusion of a monomer with a reactive group and the dye is subsequently covalently cross-linked after synthesis. In the latter case, any free dye needs to be removed by a separation step and this is usually carried out using high performance liquid chromatography (HPLC). In cases where neither dye is at the 3'-end of the probe (*e.g.* Scorpions), both dyes are incorporated as fluorophore labelled monomers. In dual labelled probe types the donor is liberated from the acceptor. If the stochiometry of the donor and acceptor fluorophore incorporation is not balanced then a low signal and/or a high background may be obtained. The quality of probes can vary greatly between suppliers and even between scales of synthesis. Quality control of such probes is important to avoid wasting time trying to optimise assays with poor quality dual labelled probes. The authors would recommend that probes should be checked following manufacture by any one or more of a number of techniques to ensure the integrity of each probe. These include mass spectrometry and/or HPLC separation using analysis of the fractions at the relevant wavelengths. Degradation and subsequent analysis utilising a spectrofluorimeter is an ideal method of determining dye stochiometry. Some suppliers offer one or more methods as a service or as part of their own quality control.

Another experimental consideration important in improving the signal:noise ratio of assays is the sequence dependent quenching of fluorescence. In particular, a fluorescein dye may be partially quenched by a neighbouring G in the sequence, which should be avoided. Whilst both dyes in a FRET pair may be fluorescent it is becoming increasingly common to find the use of non-fluorescent acceptors (quenchers). These are often termed "dark quenchers". Their use is increasing because a non-fluorescent moiety on a dual labelled probe frees the instrument's bandwidth, increasing both the utility of the instrument for multiplexing, and the dynamic range of a given detector where fluorescent emissions may have otherwise overlapped. Such quenching moieties are now commonly employed on dual labelled probes. One limitation of this approach we have observed is that it is difficult for the end user to determine the efficiency of the incorporation of these quenchers into dual labelled probes. In practice this may lead to a higher noise for the reasons described above (Zhang *et al.*, 2001). The use of gold nano-particles as efficient quenching moieties has also been reported (Dubertret *et al.*, 2001), however this has yet to be demonstrated in a homogenous PCR.

Since most instruments can excite and detect the emissions of fluorescein, a combination of fluorescein and any dye that can act as an acceptor may be used to report amplification. On most instruments the rise or drop in fluorescence from either fluorescein or the acceptor would be suitable to report amplification in various assay types. However, when a multiplexed assay is required, the instrument specification narrows the choice of dyes and assay types that can be used. Two principle approaches are used by different manufactures to allow multiplexing. Here we describe these as the universal donor method, and the universal acceptor method (see Figure 3):

1) Universal Donor. In this arrangement the system is set up to multiplex assays by monitoring the increase in acceptor emissions. Different acceptors are used to report amplification from two or more amplifications in the same tube. Typically the machine has one excitation source and detectors at longer wavelengths that are used when multiplexing *e.g.* the LightCycler®.

2) Universal Acceptor. In this arrangement the system is set up to multiplex by monitoring the increase in donor emissions that are quenched by a longer wavelength or non-fluorescent acceptor. Different donors are used to report amplification from two or more amplifications in the same tube. Typically the systems detect donors that have short wavelength emission and are all excited by the one excitation source *e.g.* the Applied Biosystems 7700 and 7000. Alternatively dyes that are spectrally well separated with longer wavelengths may be used on some instruments that have the ability to both excite at more than one wavelength, and measure emission at more than one wavelength, *e.g.* BioRad iCycler® and Cepheid SmartCycler®.

In theory most fluorometric instruments can be used to detect either universal donor or universal acceptor dye combinations, but software on most platforms is not flexible enough to allow practical application and analysis using both approaches without first exporting the data. The universal acceptor arrangement is the most common because of its effectiveness with quenched dual labelled probes (5' nuclease probes, molecular beacons etc) that are widely used. However, new instruments such as the Corbett Rotor-Gene® are flexible enough in optical design and analysis to allow the use of most, if not all fluorescent methods. It should be noted that generally the signal obtained when measuring acceptor fluorescence in universal donor methods, is of a lower value (because of energy transfer inefficiencies) and may inherently generate more noise than the universal acceptor methods, where donor fluorescence is measured. That is not to say that a comparative signal: noise is not achievable with universal donor methods. In particular this is possible where the probe and FRET efficiencies are good and the dyes are spectrally well separated, such that the donor moieties do not significantly contribute to the high background when detecting the acceptor emissions, as is achieved with the LightCycler® dye combinations (fluorescein, LC640 and LC705).

One approach to allow the universal acceptor method to be used on a system optimised for the universal donor method uses the "big dye" approach to transfer the energy of the excitation source to the reporting dye (Solinas et al., 2001). This approach has been previously applied in DNA sequencing chemistries and instruments (Ju et al., 1995). For example, on the LightCycler® the dyes for the longer wavelength detectors are only minimally excited by the blue LED. A fluorescein molecule that is chemically coupled to such a dye may now be detected in the longer wavelength detectors since the energy will be transferred by the fluorescein, in turn this energy may be transferred to a longer wavelength or dark acceptor dye. Although less efficient the method does work and these probes are available from a few suppliers. Table 2 provides a summary of commonly used dye pairs.

A passive reference is used in some systems. This dye is detected by the instrument and its signal is used to normalise for variations in volume that occur between wells and subtle differences in the optical properties of tubes and wells. These differences may be due to either volumetric errors in pipetting during preparation, and/or changes that occur during the amplification as evaporation occurs, causing a drift in the background signal. The reference dye takes no part in the reaction itself. On Applied Biosystems and similar instruments ROX (carboxy-X-rhodamine) dye is commonly used for this purpose. On instruments where there is little surface area for evaporation to occur, such as capillary systems like the Idaho and Roche LightCycler®, and in centrifugal systems where the volume and concentration is maintained in the reaction well, such as the Corbett Rotor-Gene®, a passive reference is less important.

A prerequisite of multiplexing using fluorescence is discriminating the emission signals of different fluorescent dyes at different wavelengths. Different approaches to achieve this include;

1) A fluorospectrometer that can measure a given area of the visible spectrum and use a calibration algorithm and software to deconvolute the different dye components (*e.g.* Applied Biosystems 7700).
2) A fluorimeter with discrete optical detectors using a calibration algorithm and software to deconvolute the different dye components (*e.g.* Applied Biosystems 7000 and Roche LightCycler).
3) A fluorimeter with discrete optical detectors and utilising a "factory set" calibration to deconvolute the different dye components (*e.g.* Cepheid SmartCycler®).
4) A fluorimeter with discrete optical detectors with sufficient bandwidth between dyes and narrow cut-off filter to minimise cross talk (*e.g.* Corbett Rotor-Gene®).

All approaches are equally applicable but we would recommend, for high confidence applications, the use of an algorithm to minimise the risk of false positives due to erroneous data interpretation. A good algorithm is one that completely compensates for the contribution of other dyes at the wavelength of the dye being monitored. This is usually achieved using a calibration with pure dyes. A number of factors can affect the requirement for re-calibration; these include the drift in detector sensitivities and excitation sources. Effects of specific sequences on dye emissions may also necessitate that a re-calibration be carried out to account for any shift in emission wavelength.

The 5' Nuclease Assay

The principle of a DNA synthesis dependent 5'-3' exonuclease process using the enzyme *Taq* polymerase was first described in 1989 (Gelfand,

1989). The application of this process for analysis of PCR amplification using isotopic labelling and a separation analysis was later reported (Holland *et al.*, 1991). A fluorogenic approach for monitoring the hydrolysis of HIV-1 peptides in real-time (Matayoshi *et al.*, 1990) had been described and this principle was subsequently applied to the 5' nuclease assay. A fluorogenic 5' nuclease PCR assay was first reported in 1993 (Lee *et al.*, 1993).

The 5' nuclease assay is now by far the most commonly used strand-specific probe method. This is partly because this assay was the principal detection chemistry for the first commercial real-time PCR instrument, the Applied Biosystems 7700 launched in 1996 (Bassam *et al.*, 1996). Applied Biosystems market the 5' nuclease assay for research applications under the registered TaqMan® trademark of Roche Molecular Systems Inc.

The probe is dual labelled. Typically, one dye of a FRET pair is located at the 5' end of the probe, usually the donor, and the second is labelled at or near the 3' end, usually the acceptor. However, the reverse is acceptable and more efficient quenching may be obtained by locating the acceptor closer to the donor. When dyes are coupled in close proximity the signal:noise ratio of the assay improves compared to terminally labelled probes. The optimum distance of the two dyes on the probe is dependent upon the properties of the dyes used in the FRET pair. Typically a spacing of six nucleotides is optimal. If the dyes are too close then there may be inefficient transfer of energy (Heller *et al.*, 1987) and they risk being cleaved together on the same fragment. The probe is protected at the 3' end to prevent it being extended as an amplimer by the *Taq* polymerase. A phosphate group is normally employed for this but other blocker-groups work equally well.

The 5' nuclease assay uses a probe that is designed to be cleaved by the concomitant 5'-3' exonuclease activity of the *Taq* polymerase (Holland *et al.*, 1991). This occurs whilst the *Taq* enzyme is extending the nascent 3' end of a neighbouring amplimer bound to the complementary strand. The products of hydrolysis have been shown to include 1-2 base products and it is proposed that that the polymerase displaces the first 1-2 base pairs it encounters before beginning cleavage (Holland *et al.*, 1991). The 5' nuclease assay is fundamentally different to the other strand-specific probes because the signal is generated by the 5' exonuclease process. Probe hybridisation is necessary, but not sufficient, for this to occur.

The Applied Biosystems 7700 utilises dyes that were previously optimised for Applied Biosystems sequencers, so the first commercially available 5' nuclease probes used FAM (carboxyfluorescein), JOE (carboxy-4', 5'-dichloro-2',7'- dimethoxyfluorescein), TET (carboxy-2',4,7,7'- tetrachlorofluorescein), and HEX (carboxy-2',4,4',5',7,7'-hexachlorofluorescein) as the donors, with TAMRA (carboxytetra methylrhodamine) as the universal acceptor. The ROX (carboxy-X-rhodamine) sequencing dye was used as a passive reference. In practice a large number of dye combinations is possible and a multiplex of up to six reporter dyes in one reaction has been used for SNP end-point scoring (Lee *et al.*, 1999).

A number of software tools can be used to design 5' nuclease probes. Similar instruments that followed the 7700 were developed with optics that supported dual probe systems using a universal acceptor arrangement, and a large number of oligonucleotide synthesis companies supply the dual labelled probes with the correct dyes to support the 7700 and later instruments. Applied Biosystems, who market TaqMan® probes, recommend that the design of such probes should follow these recommendations for optimum signal generation: the GC content of the probe should be in the 30-80% range, the strand with the greater number of C's rather than G's should be selected; runs of identical nucleotides should be avoided; there should be no G's at the 5' end (which would otherwise partially quench the fluorophore); the probe should have a melting temperature that is at least 5°C greater than the amplimers; and finally, the probe sequence should not overlap with the amplimer regions (Gelmini *et al.*, 1997).

The 5' nuclease assay is the only probe system where the dyes are covalently linked prior to molecular cleavage. The spacing of the dyes may be optimised with respect to Förster distance. Following cleavage, separation of donor from acceptor is maximised, although some spatial effects are observable on binding to the complementary strand. Providing the experimental parameters are optimised, the 5' nuclease system produces a signal:noise ratio that is usually much better than that achievable with other assay methods (see Figure 4). The method is very versatile and can be used on a large number of platforms. However, there are limitations and these will be outlined briefly here.

Applicability of the 5' nuclease assay is limited by the target and probe sequence requirements for efficient hydrolysis. A PCR may be shown to be highly efficient, but the signal from a hydrolysis probe may be low. Therefore design of an efficient probe is the key to an efficient PCR signal. As for most of the systems described, the assay is designed around the probe and not the amplimers *per se*. In particular, the use of this assay system for AT rich DNA can be problematic. To obtain a Tm sufficient to allow efficient cleavage of the probe it should be ~5-10 °C greater than that of the amplimers. For some sequences this would require the synthesis of very long probes that could only be synthesised with a lower yield. The development of the minor groove binding (MGB) probe technology that increases the Tm of such probes (Kutyavin *et al.*, 2000; Salmon *et al.*, 2002) has reduced this problem somewhat but adds to the complexity and cost of the probe. The difficulties associated with the synthesis of dual label probes and the signal: noise ratios obtained was discussed previously. We have observed this from almost all suppliers although this is less problematic in small-scale syntheses from reputable companies.

On many platforms the temperature transitions make up a large proportion of total assay time. On instruments with faster temperature transitions such as the Roche/Idaho LightCycler®, Corbett Rotor-Gene® and Cepheid SmartCycler®, the overall PCR assay time may be significantly reduced. Whilst the PCR process has been shown to proceed at considerable speed on fast systems, the signal:noise ratio of the 5' nuclease assay may be dramatically improved with an increase in hold time at the annealing step. The 5' nuclease process has been shown to generate signal slower than the PCR process itself (Whitcombe *et al 1999a* ; Lee *et al.*, 2002) and this is illustrated in Figure 4.

The 5' nuclease assay has been reported for a large number of applications and is supported by the majority of synthesis companies and suppliers of instruments, such that for many new users, the trademark TaqMan® is used as a synonym for real-time PCR. These applications include expression studies, detection of pathogens, allelic discrimination, and genomic applications to name but a few.

Direct Hybridisation Methods

Direct hybridisation methods use one or more probes that hybridise to the product strands during the low temperature stages of the PCR each cycle and are released from the product strand during the high temperature stages. Unlike the 5' nuclease assay the signal is not cumulative, but does increase each cycle as the amount of probe binding rises with product accumulation. Since the signal is derived from a hybridisation event the signal is proportional to the product concentration, but is also a function of the probe efficiency. The signal from hybridisation assays may be based on direct binding of product or on the competitive formation of different hybrids. Generally, where the hybridisation event is simple the effect of probe efficiency will decrease. Here we describe the different types of hybridisation probe in detail.

Linear Hybridisation Probes

A number of different probe systems fall into this class. These use the direct hybridisation of linear oligonucleotides; these are oligonucleotides that are designed to have a minimal secondary structure when not bound to target. Here we will describe the various types and use of these probes in more detail.

The most often reported probes of this class are dual hybridisation probes that were the strand-specific chemistry developed for PCR applications on the LightCycler® instrument (Cardullo *et al.*, 1988; Wittwer *et al.*, 1997). The term hybridisation probe is often used as a synonym for this probe method. This probe system uses two oligonucleotide probes each labelled with one dye of a FRET pair. Typically one dye is located at the 3' end on one oligonucleotide, and the other dye is located at the 5' end of the second oligonucleotide. The second oligonucleotide is also protected to stop primer extension using a phosphate or other blocker molecule.

These oligonucleotides are designed to hybridise adjacently on one of the natant PCR product strands such that the dyes are in a *cis* configuration with respect to the nucleic acid. Both probes are designed to bind to the non-amplimer region of one of the amplicon strands. As product accumulates both probes bind to the strand and the basis of detection is to monitor the drop in donor fluorescence, or the rise in acceptor fluorescence, or to monitor the quotient of acceptor/donor in non-multiplexed reactions. Typically fluorescence is measured at the end of the annealing step and requires 10-30 seconds to generate sufficient signal for this application.

The use of dual hybridisation probes was first described for use in PCR on the Idaho LightCycler®. The two channel optical systems of these machines used fluorescein as the donor moiety and the Amersham dye Cy5 as the acceptor. The later three channel Roche LightCycler® uses three dyes, two of which are optimised for the LightCycler®, LC640 and LC705, although it is possible to multiplex using other dyes with emission wavelengths that correspond to the discreet optical detectors of the LightCycler®. The assay is applicable on other platforms such as the Corbett Rotor-Gene® in either a universal donor arrangement or a universal acceptor arrangement by using a number of dye combinations and a non-fluorescent quenching moiety.

For applications requiring accurate quantification of initial target copy numbers a requirement is that the probes be designed to bind at approximately the same temperature, but above that of the amplimers so that efficient signal is generated during the annealing step, and melt at a temperature that will allow efficient extension by the *Taq* polymerase. This typically requires the probes to have a melting temperature of 65-75°C. If the probe Tm is too low, insufficient signal will be generated each cycle. If the probe Tm is too high, the probes may clamp the PCR and reduce amplification efficiency. The probe efficiency (the amount of signal generated by hybridisation) can be significantly reduced by probe-probe interactions if there exists any complementary sequence between probes. This is important in multiplexed reactions where the concentration and number of probes increases. Ideally the probes should be designed to bind to one strand, close to, but not overlapping the reverse amplimer region. The Tm of probes can be calculated using software tools such as the Idaho Technology applet found on their website, or using other commercial packages such as that supplied by Roche for use on the LightCycler®. If the design of probes is carried out as described then the probe efficiency will be high enough to provide accurate quantification of initial target molecules. A commonly observed effect in this and other hybridisation assays is a decrease in probe signal in later cycles. This "hook" effect is a result of increased specific product accumulation and competitive displacement of probes as the product strands anneal upon cooling (Figure 5). This effect does not reduce the ability to accurately quantify initial target numbers.

A less widely reported variation of the dual hybridisation probe is one that uses two probes with the dyes in a *trans* configuration with respect to the probed strand (Bernard *et al.*, 1998). In this embodiment one "free" probe is used in combination with a fluorescent labelled amplimer. The energy from the dye on the amplimer may be transferred to the specific probe during the low temperature steps and the probe system may be similarly applied.

One of the useful features of hybridisation probes generally is the ability to carry out a multiplex qualitative analysis utilising melting point analysis. For example the same fluorescent dye on two different probes may be used to detect two amplification species in the same reaction providing their respective melting motifs can be resolved by the instrument and software. If the software can a) determine more than one peak, and b) calculate the relative areas under these peaks using integral analysis, then a relative quantitative value for each species can be determined within a single reaction. Detection of internal controls using this approach is discussed in chapter 5.

Likewise one probe may be used to detect strand variation by designing the probe to bind over a site where strand variation may occur. For example, allelic variation is readily detected using this approach. In the case of dual hybridisation probes one probe is designed to bind to a conserved (non-variant) sequence that is adjacent to the sequence variation. This first probe is known as the "anchor" probe as it is designed to hybridise to the target over a temperature range at which the second "mutation" probes Tm will change depending on the specific sequence. The mutation probe can be observed to melt at different temperatures depending on the presence of either variant. For example, the genotype for a given sample may be determined using this approach. A single peak would indicate a homozygous genotype, and if both peaks occurred then this would be a heterozygous genotype. This is a useful method because a single probe can detect two variants, and this may be multiplexed by fluorescent emission to increase throughput if applicable. This approach can also be used to study the allele frequency in a mixed pool for a given population. For example, using the integral analysis of melting peaks described above allows a relative value for genetic variation, *e.g.* single nucleotide polymorphisms (SNPs), to be determined for a given population.

Yin-Yang probes are dual linear probes that work on the basis of competitive hybridisation between themselves and the PCR product (Li *et al.*, 2002). This principle was first demonstrated for the end-point detection of PCR products in 1989 (Morrison *et al.*, 1989). Two labelled oligonucleotides are used. A fluorophore on the reporter probe is quenched by an acceptor on a second shorter complementary probe. Upon the increase in PCR product accumulation the acceptor probe is displaced by the hybridisation of the longer target.

A limitation of these approaches is that two oligonucleotide probes are required. The application of dual hybridisation probes may be difficult for some applications where there is little or no conserved region for one or either probe. The empirical optimisation of two probes may increase the assay optimisation time. ResonSense® Hybeacons® and Light-up® probes are examples of single linear probe systems that have been described for a limited number of applications. The experimental considerations for the design and implementation of these probes will be similar to dual hybridisation probes because of their similar mode of operation. These probes will be briefly described.

ResonSense® uses a single linear probe and a DNA binding dye as a FRET pair. Signal is generated when the probe binds to the accumulating product and energy transfer occurs between the dye on the probe and the DNA binding agent in the DNA duplex (Cardullo *et al.*, 1988; Lee *et al.*, 1999: Taylor *et al* 2001). ResonSense® has been recently reported for use in real-time quantitative PCR (Lee *et al.*, 2002). The ResonSense® probe may be labelled with the fluorophore at either end or internally. A 3' probe label will prevent primer extension. If the label is at the 5' end or is internal then an additional 3' blocker molecule will be required. The probe system may be used in either a universal donor arrangement (Lee *et al.*, 2002), or in a universal acceptor arrangement using a non-fluorescent quenching moiety. For example, the minor groove binding dye SYBR® Gold may be used to transfer energy to probes labelled with Cy3, Cy5, Cy5.5, to name a few. The broad emission spectra of SYBR® Gold allows for a variety of acceptor dyes to be utilised. SYBR® Gold and related dyes have two excitation peaks, one of which is in the UV. The use of a UV excitation source would allow for efficient excitation of the SYBR® Gold whilst minimising the excitation of much longer wavelength dyes as this described. Alternatively, the drop in donor dyes such as fluorescein, can be monitored by using a longer wavelength binding dye.

Hybeacons™ are linear probes that use a fluorescent moiety attached to an internal nucleotide that exhibits enhanced fluorescence with hybridisation to specific target (French *et al.*, 2001). The probe is 3' protected to prevent primer extension during the course of the PCR. The increase in signal allows for direct quantification of initial target copy numbers and qualitative analysis of closely related sequences using melting point methodology. The simple structure of the probe and mechanism of signal generation make it ideal for multiplexing in strand variation analysis.

Light-Up™ probes (Svanik *et al.*, 2000a; Svanik *et al.*, 2000b; Isacsson *et al.*, 2000) are peptide nucleic acid probes that use a DNA binding agent to report the hybridisation of the PNA to specific DNA sequence.

The application of Light-Up™ probes for real-time PCR has been described (Wolffs *et al.*, 2001). In this probe method the probe binds to the target and brings the binding agent spatially close to the probe and this change allows it to bind with enhanced fluorescence. The dye reported in Light-Up™ probes is thiazole orange (Lee *et al.*, 1986). PNA has been shown previously to be a useful molecule in some PCRs being used to clamp amplification (Kyger *et al.*, 1998). The PNA has been shown not to clamp the PCR in Light-Up™ assays because the probe target sequence is not located near the amplimer regions.

The Molecular Beacon and Other Conformational Probes

The use of a conformational change in fluorescent labelled probes to detect nucleic acid hybridisation was first reported in 1987 (Heller *et al.*, 1987), and was later proposed for a homogeneous assay, in the context of nucleic acid amplification, in 1995 (Livak *et al.*, 1995; Young, 1996). The temperature dependent change in FRET occurring as a DNA probe, with a donor and acceptor label at either end, hybridising to its specific complement is sufficient to provide a signal (of the event) in a homogeneous assay format. The Eclipse® probe system works on this principle but utilises a minor-groove binder moiety to stabilise the structure (Afonina *et al.*, 2002). Eclipse® probes are chemically similar to the MGB probes that may be used in the 5' nuclease assay but are typically labelled with the MGB and dark acceptor on the 5' end to prevent the probe being hydrolysed. This type of probe can be used for a number of applications and melting point analysis can be used to identify sequence variation as previously described.

The molecular beacon was described first in 1995 (Tyagi *et al.*, 1996; Tyagi *et al.*, 1998) and has been reported for a number of applications including direct 16S RNA detection (Schofield *et al.*, 1997), RNA detection in Nucleic Acid Sequence Based Amplification (NASBA) (Leone *et al.*, 1998), and recently the detection of DNA binding proteins (Heyduk *et al.*, 2002).

The chemical structure of the molecular beacon may be identical to that of the probe used in the 5' nuclease assay in that it may be labelled at the 5' and 3' ends with a donor and an acceptor moiety respectively, and may also be protected from primer extension at the 3' end. However, the process of detection is different in that the probe is not designed to be consumed during the course of the reaction. The 5' and 3' sequence of the molecular beacon are designed to be complementary using a sequence not present in the target such that the two fluorescent moieties are brought close enough to allow effective energy transfer in a "hairpin" structure. The amplification is detected by the sequence dependent opening of a "loop" sequence of the probe that is complementary to the target. Therefore these probes may be monitored for certain applications using melting point methodology (Robinson *et al.*, 2000).

The difference in mode of operation of a hydrolysis probe and a probe such as a molecular beacon merits some discussion. Although the probe is chemically similar in structure, it is the process of detection that is different. However, when the process is carried out using similar reaction conditions the signal in both assays will contain a component derived from both the de-conformational change in hybridisation and a component derived from any strand specific hydrolysis by the 5'-3' exonuclease activity of the enzyme contributing to the increase in the Förster distance of the fluorphores. The contribution of either component will depend on the assay design and the enzyme used, as some enzymes have different activities and may lack the 5'-3' exonuclease activity e.g Stoffel fragment of *Taq* polymerase. The advantages and application of a molecular beacon to use both processes has recently been reported (Kong. *et al.*, 2002) and the utility of either process for SNP scoring has been discussed (Täpp *et al.*, 2000).

Typically a universal acceptor arrangement is used for molecular beacons since it can be carried out on all instruments that can analyse the 5' nuclease assay. The original work used EDANS (5-((2-aminoet hyl)amino)naphthalene-1-sulfonic acid) that is excited in the UV and emits in the blue region of the visible spectrum, and DABCYL (((4-(dimethylamino)phenyl)azo)benzoic acid) a dark acceptor. The use of a dark quencher for molecular beacons is still commonplace, however the donor moiety is more commonly one that can be used on current instruments, although any FRET dye pair will work and some have been reported (Zhang *et al.*, 2001). The use of different fluorophores for quantitative multiplexing is possible (Vet *et al.*, 1999).

A review of the use of such structures for a number of applications is given in Broude *et al.*, 2002. One application of the hairpin structure has been the development of a generic detection chemistry. This has been available in the past as Sunrise® and is now available as Amplifluor®. This chemistry uses a molecular beacon structure that is synthesised as a tail to one amplimer (Nazarenko *et al.*, 1997). During PCR the *Taq* polymerase polymerises the complement sequence so that upon hybridisation the hairpin structure does not form, and the donor and acceptor are separated as the basis of detection. A further simplified embodiment uses a unique sequence that is incorporated into the tail of one amplimer called a "z" sequence. A generic Amplifluor® probe called a UniPrimer® can initiate primer extension since it has specificity for the z-sequence. When the complement is synthesised in the next cycle of PCR then the two dyes are separated as previously described.

Whilst Amplifluor® is a generic detection approach, the probe having no specificity for the internal sequence of the amplicon, it does offer some advantages over DNA binding approaches for detection. Firstly the signal:noise ratio of a FRET based assays such as this is typically better than that obtained using a DNA binding dye, although non-specific priming may still generate signal. Secondly the probe may be a common probe sequence. A similar approach using a primer incorporated sequence for a generic approach utilising common hydrolysis probes has also been reported for single tube genotyping (Whitcombe *et al.*, 1998). Finally this approach allows for a generic approach to multiplexing using different fluorophore emissions.

Self-Probing Amplicons

A self-probing amplicon is an amplicon that has a probe chemically incorporated into the PCR product. The principle was first reported in 1999 and used a molecular beacon that was connected to the 5' end of an amplimer via a linker molecule (Newton *et al.*, 1993;

Whitcombe *et al.*, 1999a; Whitcombe *et al.*, 1999b). This probe is an oligonucleotide containing a sequence that is complementary to a region synthesised by the extension of the amplimer it is attached to. The probe sequence also contains stem sequences at either end that hybridise and bring two fluorphores in a FRET pair spatially close together. The linker prevents the enzyme generating the complement of the probe during the extension of the reverse strand. Following denaturation of the duplex at the high temperature step of PCR, the probe binds to its complementary sequence that now forms part of the same molecule. The energetics favour this structure rather than that of the complementary stem structures within the probe sequence. The binding of the probe sequence to target increases the distance between dye pairs allowing signal generation. The self-probing unimolecular mechanism has been proven to be the dominant mechanism of signal generation, rather the bimolecular mechanism of the probe binding to other amplicons (Thelwell *et al.*, 2000). It has also been shown that the cleavage of such probes does not occur due to the 5'-3' exonuclease activity of the enzyme (Thelwell *et al.*, 2000). In the original work the probe used fluorescein as the donor moiety and methyl red as a non-fluorescent acceptor. These probes are known commercially as Scorpion® probes.

The main advantage of this method is that the probe is made in the same synthesis as the amplimer and therefore a separate probe is not required. The speed of signal generation has been reported to be much greater than that of the respective native molecular beacon utilising bimolecular probing (Whitcombe *et al.*, 1999a; Thelwell *et al.*, 2000), and actually more efficient when using short annealing times (Thelwell *et al.*, 2000). Since the probe becomes linked to the sequence it is probing, the event is in effect a zero order reaction (Saha *et al.*, 2001). The equal stochiometry of the probe to target means that the amplification signal is directly proportional to the amount of initial target species. This has also been illustrated utilising a ResonSense® probe attached to an amplimer (Lee *et al.*, 2002). This type of probe has been termed Angler®. These types of assays will be of most use in improving throughput on faster thermal cyclers.

Other embodiments of the Scorpions® assay include the duplex Scorpion primer method (Solinas *et al.*, 2001). In this method the amplimer is labelled with a linear probe that has specificity for the strand the amplimer will generate. This probe contains one fluorophore of a FRET pair. A second probe that is the complement of the probe is free in solution. This probe is labelled with the second fluorophore in the FRET pair. Prior to the priming by the amplimer the probes interact and FRET occurs. When the amplimer extends the first probe is able to bind to the newly generated complement in an intra-molecular fashion. The first probe is then spatially separated from the second and the change in FRET indicates amplification.

IntraTaq™ is the self-probing embodiment of the 5' nuclease assay that has been shown to be more efficient in terms of the signal:noise ratio and the speed of the reaction (Solinas, *et al.*, 2002). In theory it should be possible to link any probe system to the amplimer in a similar fashion, and one would expect a commensurate improvement in signal with respect to speed for assays based on hybridisation. One advantage of the self-probing amplicon that is not immediately apparent is that the probe is better controlled for, compared to a separate probe, since it is synthesised on the same molecule (Lee *et al.*, 2002). The main disadvantage of such probes are that the synthesis is more complex and therefore more costly, and the availability is limited to those oligonucleotide synthesis companies that can reproduce these chemistries.

Summary and Future Improvements

The chemistries described can all be applied effectively to carry out the real-time detection of PCR amplification. The effectiveness of a chemistry depends on the instrument being used, the suitability of the software for correct data analysis, and both the target sequence and application. The specifications of new instruments are improving and the development of new PCR real-time detection using methods such as fluorescence polarisation, time resolved fluorescence and application of quantum dot technology are possible. These should allow for improvements in both the level of multiplexing and the signal: noise ratios obtainable using current approaches. Such methods could improve the speed, robustness, and costs associated with real-time PCR.

Acknowledgements

The authors would like to thank the editors for the invitation to contribute to this text. The work of the authors described in this text was funded by the UK Ministry of Defence.

References

Afonina, I.A., Reed, M.W., Lusby, E., Shishkina, I.G., and Belousov, Y.S. 2002. Minor Groove Binder-Conjugated DNA probes for quantitative DNA detection by hybridisation-triggered fluorescence. BioTechniques. 32: 940-949.

Bassam, B.J., Allen, T., Flood, S., Stevens, J., Wyatt, P., and Livak, K.J. 1996. Nucleic acid sequence detection systems: Revolutionary automation for monitoring and reporting PCR products. Australasian Biotechnology. 6: 285-294.

Becker A., Reith A., Napiwotzki, J., and Kadenbach B. 1996. A quantitative method of determining initial amounts of DNA by polymerase chain reaction cycle titration using digital imaging and a novel DNA stain. Anal. Biochem. 237: 204-207.

Bernard, P.S., Lay, M. J., and Wittwer C.T. 1998. Integrated amplification and detection of the C677T point mutation in the methylenetrahydrofolate reductase gene by Fluoresence Resonance Energy Transfer and probe melting curves. Anal. Biochem. 255: 101-107.

Broude, N.E. 2002. Stem-loop oligonucleotides: a robust tool for molecular biology and biotechnology. Trends in Biotechnology. 20: 249-256.

Cardullo, R.A., Agrawal, S., Flores, C., Zamecnik, P.C., and Wolf, D.E. 1988. Detection of Nucleic Acid hybridisation by non-radiative fluorescence resonance energy transfer. Proc. Natl. Acad. Sci. USA.

85:8790-8794.

Clegg, R.M. 1992. Fluorescence Resonance Energy Transfer and Nucleic Acids. In: Methods in Enzmology. D.M.J. Lilley and J.E. Dahlber, Eds. Academic Press. New York. 211: 353-388.

Dubertret, B., Calame, M., and Libchaber, A. 2001. Single-mismatch detection using gold-quenched fluorescent oligonucleotides. Nat. Bio. 19: 365-370.

French, D.J., Archard, C.L., Brown, T., and McDowell, D.G. 2001. Hybeacon™ probes: a new tool for DNA sequence detection and allele discrimination. Mol. Cell. Probes. 15: 363-374.

Gelfand, D.H. 1989. Taq DNA polymerase. In PCR Theory, principles, and application to DNA amplification. 1988. In Erlich, H. A. (ed.) *PCR Technology*. Stockton Press, NY. p. 17-22.

Gelmini, S., Orlando, C., Sestini, R., Vona, G., Pinzani, P., Ruocco, L., and Pazzagli, M. 1997. Quantitative polymerase chain reaction-based homogeneous assay with fluorogenic probes to measure c-erB-2 oncogene amplification. Clin. Chem. 43: 752-758.

Heller, M.J., and Jablonski, E.J. 1987. Fluorescent stokes shift probes for polynucleotide hybridisation assays. European patent application 86116652.8.

Heyduk, T., and Heyduk E. 2002. Molecular beacons for detecting DNA binding proteins. Nat Biotech. 20: 171-176.

Higuchi, R., Dollinger, G., Walsh, P.S., and Griffith, R. 1992. Simultaneous amplification and detection of specific DNA sequences. BioTechnology. 10: 413-417

Higuchi, R., Fockler, C., Dollinger, G. and Watson , R. 1993. Kinetic PCR analysis: Real-time monitoring of DNA amplification reactions. BioTechnology. 11: 1026-1030.

Holland, P.M., Abramson, R.D., Watson, R., and Gelfand, D.H. 1991. Detection of specific polymerase chain reaction product by utilising the 5'-3' exonuclease activity of *Thermus aquaticus* DNA polymerase. Proc. Natl. Acad. Sci. USA. 88: 7276-7280.

Isacsson, J., Cao, H., Ohlsson, L., Nordgren, S., Svanvik, N., Westman, G., Kubista, M., Sjöback, R. and Sehlstedt, U. 2000. Rapid and specific detection of PCR products using light-up probes. Mol. Cell. Probes. 14: 321-328.

Ju, J., Ruan, C., Fuller, C.W.,Glazer, A.A., and Mathies, R. A. 1995. Fluorescence energy transfer dye-labeled primers for DNA sequencing and analysis. Proc. Natl. Acad. Sci. USA. 92: 4347-4351

Kong, D.M, Gu, L., Shen H. X., and Mi, H.F. 2002. A modified molecular beacon combining the properties of TaqMan probe. Chem Commun. 8: 854-855.

Koshkin, A.A., Rajwanshi, V.K. and Wengel, J. 1998. Novel convenient syntheses of LNA [2.2.1] bicyclo nucleosides. Tetrahedron Lett. 39: 4381-4384.

Kutyavin, I. V., Afonina, I. A., Mills, A, Gorn, V.V., Lukhtanov, E.A., Belousov, E.S., Singer, M.J., Walburger, D.K., Lokhov S.G., Gall, A.A., Dempcy, R., Reed, M.W., Meyer, R.B., Hedgpeth, J. 2000. 3' minor groove binder-DNA probes increase sequence specificity at PCR extension temperatures. Nuc. Acid. Res. 28: 655-661.

Kyger, E.M., Krevolin, M.D., and Powell, M.J. 1998. Detection of the hereditary hemochromatosis gene mutation by real-time fluorescence polymerase chain reaction and peptide nucleic acid clamping. Anal. Biochem. 260 :142-148.

Lee, L.G., Chen, C.H., and Chiu, L.A. 1986. Thiazole orange: a new dye for retiulocyte analysis. Cytometry. 7: 508-517.

Lee, L.G., Connell, C.R., and Bloch, W. 1993. Allelic discrimination by nick-translation PCR with fluorogenic probes. Nuc. Acid. Res. 21: 3761-3766.

Lee, M.A., Brightwell, G, Leslie, D., Bird, H., and Hamilton, A. 1999. Fluorescent detection techniques for real-time multiplex strand specific detection of *Bacillus anthracis* using rapid PCR. J. Applied. Micro. 87: 218-223.

Lee, L.G, Livak, K.J., Mullah, B., Graham, R.J., Vinayak, R.S. and Woudenberg T.M. 1999. Seven-colour, homogeneous detection of six PCR products. Biotechniques. 27: 342-349.

Lee, M.A., Siddle, S.L. and Hunter, R.P. 2002. ResonSense®: Simple linear probes for quantitative homogeneous rapid polymerase chain reaction. Anal. Chimic. Acta. 457: 61-70.

Leone, G., Schijndel, H., Gemen, B., Kramer, F.R. and Schoen, C D. 1998. Molecular beacon probes combined with amplification by NASBA enable homogenous, real-time detection of RNA. Nuc. Acid. Res. 26: 2150-2155.

Li, Q., Luan, G, Guo, Q., and Liang, J. 2002. A new class of homogeneous nucleic acid probes based on specific displacement hybridisation. Nuc. Acid. Res. 30: e5.

Livak, K.J., Flood, S.A.J., Marmaro, J., Giusti, W., and Deetz, K. 1995. Oligonucleotides with fluorescent dyes at opposite ends provide a quenched probe system useful for detecting PCR product and nucleic acid hybridisation. PCR. Meth. Appl. 4: 357-362.

Matayoshi, E.D., Wang, G.T., Krafft, G.A., and Erickson, J. 1990. Novel fluorogenic substrates for assaying retroviral proteases by resonance energy transfer. Science. 247: 954-958.

Morrison, L.E., Halder, T.C., and Stols, L.M. 1989. Solution-phase detection of polynucleotides using interacting fluorescent labels and competitive hybridisation. Anal Biochem. 183: 231-244.

Morrison, T.B., Weis, J.J., and Wittwer, C.T. 1998. Quantification of low-copy transcripts by continuous SYBR®Green 1 monitoring during amplification. BioTechniques. 24:6 : 954-962.

Mullis, K.B., and Faloona, F.A. 1987. Specific synthesis of DNA *in vitro* via a polymerase catalyzed chain reaction. Meth. Enzymol. 155: 335-350.

Nazarenko, I.A., Bhatnagar, S.K., and Hohman, R.J. 1997. A closed tube format for amplification and detection of DNA based on energy transfer. Nuc. Acid. Res. 25(12): 2516-2512.

Newton, C. R., Graham, A., Heptinstall, L.E., Powell, S.J, Summers, C., Kalsheker, N., Smith, J.C., and Markham, A.F. 1989. Analysis of any point mutation in DNA. The amplification refractory mutation system (ARMS). Nuc. Acids. Res. 17: 2503-2516.

Newton, C.R., Holland D., Heptinstall, L.E., Hodgson, I., Edge, M.D., Markham, A.F., and Mclean, M.J. 1993. The production of PCR products with 5' single-stranded tails using primers that incorporate novel phosphoramidite intermediates. Nuc. Acid. Res. 21: 1155-1162.

Ogura, M., and Mitsuhahi, M. 1994. Screening Method for a Large Quantity of Polymerase Chain Reaction Products by Measuring YOYO-1 Fluorescence on 96-Well Polypropylene Plates. Anal. Biochem. 218: 458-459.

Ortiz, E., Estrada, G. and Lizardi, P.M. 1998. PNA molecular beacons for rapid detection of PCR amplicons. Mol. Cell. Probes. 12: 219-226.

Ririe, K. M., Rasmussen, P.R., and Wittwer, C. T. 1997. Product differentiation by analysis of DNA melting curves during the polymerase chain reaction. Anal. Biochem. 245: 154-160.

Robinson, J.K., Mueller, R., Filippone, L. 2000. New molecular beacon technology. Am. Lab. 32:30-24.

Saha, B. K., Tian, B., Bucy, R.P. 2001. Quantitation of HIV-1 by real-time PCR with a unique fluorogenic probe. J. Virol. Meth. 93: 33-42.

Saiki, R.K., Scharf, S., Faloona, F.A., Mullis, K.B., Horn, G.T., Erlich, H.A., and Arnheim, N. 1985. Enzymatic amplification of beta-globin genomic sequences and restriction site analysis for diagnosis of sickle cell anemia. Science. 230: 1350-1354.

Salmon, M. A., Vendrame, M., Kummert, J., and Lepoivre, P. 2002. Detection of apple chlorotic leaf spot virus using a 5' nuclease assay with a fluorescent 3' minor groove-DNA binding probe. J. Virol. Meth. 104: 99-106.

Schofield, P., Pell, A.N., and Krause, D.O. 1997. Molecular Beacons: Trial of a fluorescence-based solution hybrization technique for ecological studies with ruminal bacterial. App. Env. Micro. 63: 1143-1147.

Selvin, P.R. 1995. Fluorescence Resonance Energy Transfer. Meth. Enzymol. 246: 300-334.

Solinas, A., Brown, L.J., McKeen, C., Mellor, J.M., Nicol, J.T.G, Thelwell, N., and Brown, T. 2001. Duplex scorpion primers in SNP analysis and FRET applications. Nucleic Acids Research. 29: E96

Solinas, A., Thelwell, N., and Brown, T. 2002. Intramolecular TaqMan probes for genetic analysis. Chem. Commun. 2272-2273

Svanik, N., Westman, G., Wang, D., and Kubista, M. 2000a. Light-Up probes: Thiazole orange-conjugated peptide nucleic acid for detection of target nucleic acid in homogeneous solution. Anal. Biochem. 281: 26-35.

Svanik, N., Nygren, J. Westman, G., and Kubista, M. 2000b. Free-probe fluorescence of light-up probes. J. Am. Chem. Soc. 123: 803-809.

Taylor, M.J., Hughes, M.S. Skuce, R.A., and Neill, S.D. 2001. Detection of *Mycobacterium bovis* in bovine clinical specimens using real-time fluorescence and fluorescence energy transfer probe rapid-cycle PCR. J. Clinic. Micro. 39: 1272-1278.

Täpp, I., Malmberg, L., Rennel, E., Wik, M., and Syvänen, A.-C. 2000. Homogeneous scoring of single nucleotide polymorphisms: comparison of the 5'- nuclease TaqMan® assay and molecular beacon probes. Biotechniques. 28: 732-738.

Thelwell, N., Millington, S., Solinas, A., Booth, J and Brown, T. 2000. Mode of action and application of scorpion primers to mutation detection. Nuc. Acid. Res. 28: 3752-3761.

Tyagi, S., and Kramer, F.R. 1996. Molecular Beacons: probes that fluoresce upon hybridisation. Nat. Biotech. 14: 303-308.

Tyagi, S., Bratu, D.P., and Kramer, F.R. 1998. Multicolour molecular beacons for allele discrimination. Nature Biotech. 16:49-53.

Vet, J.A.M., Majithia, A.R., Marras, S.A.E, Tyagi, S., Dube, S., Poiesz, B.J., and Kramer, F.R. 1999. Multiplex detection of four pathogenic retroviruses using molecular beacons. Proc. Nat. Acad. Sci. USA. 96: 6394-6399.

Whitcombe, D., Brownie, J., Gillard, H.L., Mckechnie, D., Theaker, J., Newton, C.R., and Little, S. 1998. A homogeneous fluorescence assay for real-time PCR amplicons: its application to real-time, single tube genotyping. Clin. Chem. 44: 918-923.

Whitcombe, D., Theaker, J., Guy, S.P., Brown, T., and Little, S. 1999a. Detection of PCR products using self-probing amplicons and fluorescence. Nat. Biotech. 17: 804-807.

Whitcombe, D., Kelly, S., Mann, J., Theaker, J., Jones, C., and Little, S. 1999b. Am. J. Hum. Genet. 65: 2333.

Wittwer, C.T., Ririe, K.M., Andrew, R.V., David, D.A., Grundy, R.A., and Balis, U.J. 1997. The LightCycler™: a microvolume multisample fluorimeter with rapid temperature control. BioTechniques. 22:176-181.

Wolffs, P., Knutsson, R., Sjöback, R., and Rådström, P. 2001. PNA-based light-up probes for real-time detection of sequence-specific PCR products. BioTechniques 31:766-771.

Young, D. 1996. Viral load quantitation:- An integral part of future health management. Australasian Biotechnology. 6: 295.

Zhang, P., Beck, T., and Tan, W. 2001. Design of a molecular beacon DNA probe with two fluorophores. Angew. Chem. Int. 40: 402-405.

4

Performing Real-Time PCR

K. J. Edwards

Abstract

Optimisation of the reagents used to perform PCR is critical for reliable and reproducible results. As with any PCR initial time spent on optimisation of a real-time assay will be beneficial in the long run. Specificity, sensitivity, efficiency and reproducibility are the important criteria to consider when optimising an assay and these can be altered by changes in the primer concentration, probe concentration, cycling conditions and buffer composition. An optimised real-time PCR assay will display no test-to-test variation in the crossing threshold or crossing point and only minimal variation in the amount of fluorescence. The analysis of the real-time PCR results is also an important consideration and this differs from the analysis of conventional block-based thermal cycling. Real-time PCR provides information on the cycle at which amplification occurs and on some platforms the melting temperature of the amplicon or probe can be determined.

Optimisation of Real-Time PCR Assays

Real-time PCR assays require optimisation in order that robust assays are developed which are not affected by normal variations in the target

DNA, primer or probe compositions. A robust assay is defined as an assay in which these 'normal' variations cause no effect on the crossing threshold (CT) also known as crossing point (CP) and have only a minimal effect on the observed amount of fluorescence. The important criteria for optimisation are specificity, sensitivity, efficiency and reproducibility. It is important to decide before commencing optimisation which type of assay is required for a particular application. For example, there is little point in developing a quantitative assay if a simple qualitative assay will be just as informative. Melting curve analysis may also be required for product differentiation and in this case should be considered at the planning stage. If the real-time assay is based on conversion of an existing block-based assay it is important to note that the cycling conditions used for conventional block-based thermal cycling may not always translate easily to a real-time format and so it is important to consider re-optimisation of the assay (Teo *et al.*, 2001).

The same principles of optimisation apply to assays run on all real-time platforms. The following criteria should be optimised: buffer composition, cycle conditions, magnesium chloride ($MgCl_2$) concentration, primer concentration, probe concentration and template concentration. Commercial master mixes, which are widely available, simplify the optimisation and are convenient. In this chapter all aspects of optimisation and analysis of real-time PCR results will be discussed.

PCR Master Mix

Commercial master mixes are available for most of the real-time platforms and although some are marketed for specific instruments and probe formats, they often work equally well on other instruments. Mixes are provided in easy-to-use formats and often contain additional features such as use of dUTP allowing the enzyme uracil-DNA glycosylase to be used to prevent cross-over contamination. Some commercial master mixes contain the uracil-DNA glycosylase as well as the dUTP and carry-over contamination never needs to be considered. However, it is important to note that the use of dUTP has been shown to decrease PCR sensitivity. There is little doubt that the use of commercial master mixes can simplify optimisation and promotes uniformity of assay performance over time. However, for some laboratories the cost cannot be justified.

Preparation of in-house master mixes can be effective and assays can perform well over time if the reaction components are well optimised. The master mix requires thermostable polymerase, buffer, dNTPs and $MgCl_2$ though the latter will be discussed later in the chapter. The correct amount of polymerase is essential and there can often be a narrow optimal concentration range. Not enough polymerase leads to inefficient amplification, low fluorescence and loss of sensitivity and leads to a high CT value. Too much enzyme also leads to low fluorescence and can contribute to the production of primer dimers and production of other non-specific amplicons. The composition of the core buffer can be essential for some real-time platforms and can affect the Tm of the primers/probes and the performance of the enzyme. For example, when using glass capillaries it is essential that a protein such as BSA is included in the buffer to prevent the DNA from binding to the glass. The optimal concentration of dNTPs is usually wide and the CT values are not affected by 2-4 x increases or decreases, however, the amount of fluorescent signal can be affected. An important consideration is the use of dUTP in the master mix. *Taq* polymerase preferentially incorporates dATP, dTTP, dGTP or dCTP and although it can incorporate dUTP this is usually less efficient. However, the use of dUTP may be considered essential to prevent carry over contamination and a balance has to be found between sensitivity and the consequences of contamination. Differences of up to two cycles can be observed between master mixes containing dUTP and dTTP.

When the assay produces a large quantity of primer dimer or if the sensitivity of the assay is low, a hot-start technique can be applied to increase the stringency of the reaction (D' Aquila *et al.*, 1991; Chou *et al.*, 1992). Hot-start PCR can be achieved by using a commercially available master mix with an in-built hot-start or by adding commercially available anti-*Taq* DNA polymerase antibodies. When the PCR is performed in glass capillaries, *i.e.* using the LightCycler, the use of hot-start has been shown to improve performance as it is thought that the binding of *Taq* polymerase and magnesium ions to glass is reduced due to the hot-start mechanism (Teo *et al.*, 2002). Antibody based hot-start methods can be advantageous as the recommended activation time is only 1-3 minutes compared to 10 minutes for chemically modified *Taq* polymerase. This time difference can be important where fast PCR results are required.

Regardless of whether commercial master mix or an in-house preparation is used the following components (primers, probes, $MgCl_2$ and template concentration) require optimisation.

Primers

Primers should be designed with the aid of appropriate software and attention should be paid to the optimal size of the amplicon produced in the real-time application. Real-time PCR products are usually <500 bp, this may be much smaller than the amplicon size generated on a block-based thermal cycler. The following guidelines should be observed for optimal and accurate primer design:

- The primer 3' ends should be free from secondary structure, repetitive sequences, palindromes and highly degenerate sequences.
- Forward and reverse primers should not have significant complementary sequences.
- The forward and reverse primers should have equal GC contents, ideally between 40-70%.
- The binding sites should not have extensive secondary structure.

The annealing temperature as determined using primer design software should be used as an initial guide, however, it is important to note that this can often vary greatly from the experimentally determined annealing temperature. One approach, which can be used to determine the true annealing temperature is to synthesize the complementary primer sequence, hybridise the two primers and perform a melting analysis that allows the Tm to be observed under the conditions that will be used in the PCR reaction. The cost of synthesising another oligonucleotide is often justified by the reduction in time spent repeating the experiment at different annealing temperatures (Teo *et al.*, 2002).

Time should also be spent determining the optimal primer concentration and it is best to try a range of final concentrations from 0.1 μM to 0.5 μM for both primers. For some assays the optimal concentration for the two primers will not be the same, for example, to achieve good melting curves it is sometimes necessary to use asymmetric PCR, where a lower concentration of one primer is used, to increase the amount of the strand complementary to the probe (Lyon *et al.*, 1998; Phillips *et al.*, 2000). This improves the melting curve because increased amplification of the target strand can reduce the competition between the template strands and template-probe hybridisations. Some manufacturers recommend performing a primer optimisation matrix as outlined in Table 1. The optimal combination is the one that gives the lowest CT value (Figure 1).

Fluorescent Probes

As with the primers, real-time PCR probes are best designed using dedicated software such as Primer Express (Applied Biosystems) for hydrolysis probes, hybridisation probe design software from Roche Applied Science or Beacon Designer from Bio-Rad. For any of the probe formats the annealing temperature should be higher than that of the primers in order to ensure that the primers are binding and

Table 1. Example of a primer optimisation matrix.

Reverse/Forward	50nM	300nM	500nM	900nM
50nM	*50/50*	*300/50*	*500/50*	*900/50*
300nM	*50/300*	*300/300*	*500/300*	*900/300*
500nM	*50/500*	*300/500*	*500/500*	*900/500*
900nM	*50/900*	*300/900*	*500/900*	*900/900*

synthesising the product before the probe is able to participate in the reaction. It is important that the probes are not able to act in the PCR as primers and to prevent this either a fluorescent molecule or a blocker, for example phosphate, should be placed at the 3' end of the sequence. Probes complementary to the 3' termini of the PCR primers should be avoided since they may hybridise to the primers. Primer elongation may then occur leading to primer-probe dimers that affect amplification efficiencies (Roche Molecular Biochemicals). As with the primers it is important to try a range of probe concentrations to determine which is optimal. A wide range of probe concentrations is possible but if the concentration is too low no fluorescent signal will be observed and if the concentration is too high it can lead to a high fluorescent background. The probe concentration should be optimised after optimising the primer concentration and as with the primer concentration, a range of probe concentrations should be tried between 0.1 uM and 0.5 uM and the concentration which gives the lowest CT values and the highest fluorescent signal should be selected (Figure 2).

SYBR Green I

If the DNA binding dye SYBR Green I is used in the fluorescent detection system it should be optimised for each set of PCR primers. As the amount of SYBR Green I is increased an increase in the amount of fluorescence can be observed. The recommended concentration is a 1:10,000 dilution of the neat stock as supplied by the manufacturer. High concentrations of SYBR Green I have been shown to adversely affect the PCR amplification as they inhibit *Taq* polymerase (Ririe *et al.*, 1997; Wittwer *et al.*, 1997). DNA binding dyes also influence melting temperatures and consequently it is important to use only the optimised level. Some researchers have suggested that SYBR Gold is preferable to SYBR Green I due to its stability during long-term storage (*Lee et al.*, 1999).

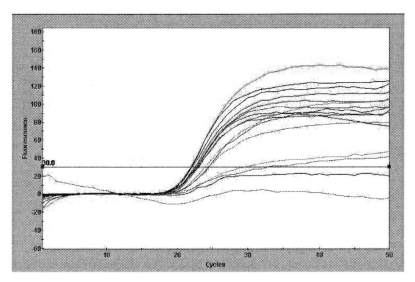

Figure 1. Primer optimisation experiment. The same sample was analysed at 12 different forward and reverse primer concentrations as detailed in Table 1. This experiment was run on the Smart Cycler using the same concentration of hydrolysis probe in each tube. The optimal primer concentration is the one that gives the lowest CT value.

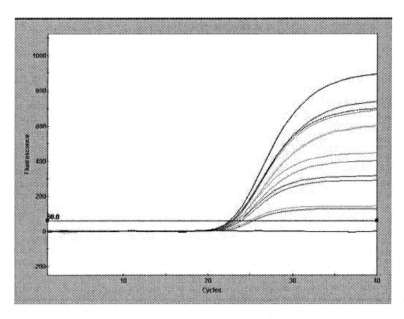

Figure 2. Probe optimisation experiment. Using two different primer concentrations, five different hydrolysis probe concentrations (250 nM, 200 nM, 150 nM, 100 nM and 50 nM) were run on the Smart Cycler. The optimal probe concentration is the one that gives the lowest CT value and the highest fluorescent signal.

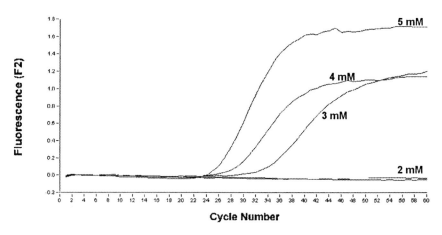

Figure 3. MgCl₂ optimisation experiment. The same sample was analysed at four different MgCl₂ concentrations (5 mM, 4 mM, 3 mM, and 2 mM) on the LightCycler. Using 5 mM MgCl₂ gave the lowest CP value and the highest fluorescent signal and was therefore the optimal concentration.

Magnesium Chloride

A key variable is the magnesium chloride ($MgCl_2$) concentration since Mg^{2+} ions are known to affect both the specificity and the yield of PCR (Oste, 1988). Concentrations that are too low or too high can have deleterious effects on PCR, high concentrations may lead to incomplete denaturation and low yields, whereas levels that are too low reduce the ability of polymerase to extend the primers. High $MgCl_2$ levels also lead to increased production of non-specific products and primer artifacts including primer dimers. This should always be avoided but especially in sensitive quantitative assays. It is recommended that a $MgCl_2$ titration is performed for each primer/probe set. This work is usually best performed after completing optimisation of primer and probe concentration. The optimal $MgCl_2$ concentration will generally be the lowest amount that gives the minimum CT value, the highest fluorescent intensity and the steepest slope of the fluorescence/cycle plot (Figure 3). The optimal $MgCl_2$ concentration for DNA assays is usually between 2-5 mM and for RNA is usually between 4-8 mM. Commercially available real-time PCR buffers are now available where

MgCl$_2$ optimisation is not required for example, QuanTitech (Qiagen, Crawley, UK) or FastStart Plus (Roche Diagnostics, Lewes, UK).

Template Purity and Concentration for Optimisation

When optimising assays it is advisable to use a minimum of a high and a low template concentration as too much DNA can inhibit the reaction and too little may be undetectable. For some assays it may not be necessary to use purified DNA or RNA however, if the primers are not amplifying it is worth repeating the PCR with pure template in the event that PCR inhibitors are present. If the CT is <10 cycles then a higher dilution should be prepared or if the CT is >30 then higher concentrations should be used. If a serial dilution series of template is prepared and amplified then an optimised assay should display an equal number of cycles between amplification of each dilution.

Cycling Conditions

The use of optimal cycling conditions are essential for real-time as with any PCR reactions (Rasmussen, 1992). The initial cycling conditions recommended for both hybridization probes and hydrolysis probes are given in Table 2. The optimal annealing temperature should be determined as described in the primer section and the optimal extension time should be calculated using the following formula: extension time (s) = amplicon length (bp) ÷ 25. Note: when hot-start techniques are employed the initial denaturation time may need to be increased. If Primer Express is used to design hydrolysis probes then they should all work optimally under the same conditions (Table 2).

PCR Controls

No template (negative) controls should always be included in real-time runs and where possible appropriate positive controls should also be included. When the assay is quantitative standards of known concentration should be included, although on some platforms it

Table 2. Cycling conditions

	Hybridisation Probe (performed on Roche LightCycler)	Hydrolysis Probe (performed on ABI Sequence Detection System)
Initial Denaturation	2-10 min @ 95°C	10 min @ 95°C
Denaturation	0 s @ 95°C	15 s @ 95°C
Annealing	5 s @ annealing temperature	60 s @ 60°C
Extension	Optimal time dependent on amplicon length @72°C	
Ramp Rate	20°C/s or reduce to 2-5°C if annealing temperature <55°C	
Fluorescent Acquisition	At end of annealing step	At end of annealing step
Melting Curve	30 s @ 55°CRamp rate 0.1°C/s with acquisition mode on continuous temperature increased to 80°C	

is possible to perform quantitative analysis using a standard curve generated in a previous run thus maximising the number of samples which can be analysed. Where melting curve analysis is used for product differentiation then it is advisable to include samples that will melt at each of the possible melting temperatures (Logan *et al.*, 2000; Edwards *et al.*, 2001). This allows for any drift in melting temperature that may be observed over time or with different batches of fluorescent probe.

To control for PCR inhibitors an additional inhibition control template should be included in each sample. This inhibition control should be amplified by the same primers as the experimental target but should contain a different probe binding site. When the reaction is significantly affected by inhibitors, their presence is indicated by the failure of both control and experimental target amplification, or by low amplification efficiency. Inhibition may be minimised by diluting the sample, by using an alternative purification technique or by obtaining a different sample. If an inhibition control is not available, negative samples should be analysed for a different target or spiked with a positive control and re-amplified. The use of inhibition controls is particularly important when real-time PCR is being used for clinical diagnostics or other applications where a false negative result could be critical.

ROX Passive Reference

Real-time platforms that use peltier blocks as the heating/cooling mechanism require the use of a reference signal to normalize the fluorescent signal across the block. By using the ROX passive reference all fluorescent signals are normalized leading to more accurate and reproducible results. When the ROX passive reference is used it is important to note that everything in the real-time PCR is relative, which means that if the ROX passive reference changes then the baseline, CT value and amount of fluorescence will also change. When optimising the passive ROX reference there is a narrow optimal range. If the amount is too high then background noise will be reduced but a low signal will be hard to distinguish from the background and this may lead to false negative results. If the ROX level is set too low the background noise is increased but a strong signal will be easy to distinguish from the background for high copy number samples, however, weak signals may be lost in the background.

Analysis of Real-Time PCR Results

Real-time PCR results can be analysed in a variety of ways depending on the application. Analyses of quantitative standards can be used to generate PCR curves and from these the cycle number at which the fluorescent signal increases above the background fluorescence can be determined. This cycle number can be used for comparison of results from run-to-run and can be used to generate qualitative positive/negative results. In quantitative assays the standard curve is used to determine the copy number of unknown samples and again this can be compared with other samples. Real-time PCR reactions can also be analysed by melting curve determination when methods including hybridisation probes and SYBR Green I intercalation have been used. Melting curves are useful for differentiating primer dimers from specific PCR products (primer dimers usually melt at lower temperatures than specific PCR products) or for differentiating different PCR products in mutation detection. During the initial optimization of real-time PCR assays it is useful to analyse the results by agarose gel electrophoresis in order to correlate product length with melting peaks and to identify any primer artifacts.

Conclusions

Real-time PCR results are dependent on the optimal concentration of each of the reaction constituents as well as on the interrelations of each of these components. The use of commercially available master mixes, which have already been optimised, provides a solid basis for a robust assay. Optimisation of $MgCl_2$, primer and probe concentration is still required to obtain maximum sensitivity and PCR efficiency but can be performed relatively easily. Depending on the platform and probe format employed some optimisation of the cycling conditions may be required. When using a well optimised assay the target will reliably amplify at the same cycle number although variations may be observed in the amount of fluorescence observed as this is more dependent on not only the PCR efficiency but also on variations in the availability of each of the different reaction components.

References

Chou, Q., Russell, M., Birch, D.E., Raymond, J. and Bloch, W. 1992. Prevention of pre-PCR mis-priming and primer dimerization improves low-copy-number amplifications. Nucleic Acid Res. 20: 1717-1723.

D'Aquilia, R.T., Bechtel, L.J., Videler, J.A., Eron J.J., Gocczyca, P. and Kaplan J.C. 1991. Maximizing sensitivity and specificity of PCR by pre-amplification heating. Nucleic Acid Res. 19: 3749.

Edwards, K.J., Kaufmann, M.E., and Saunders, N.A. 2001. Rapid and accurate identification of coagulase-negative staphylococci by real-time PCR. J. Clin. Microbiol. 39: 3047-3051.

Lee, M.A., Brightwell, G., Leslie D., Bird, H. and Hamilton A. 1999. Fluorescent detection techniques for real-time multiplex strand specific detection of *Bacillus anthracis* using rapid PCR. J. Appl. Microbiol. 87: 218-233.

Logan, J.M., Edwards, K.J., Saunders, N.A., and Stanley, J. 2001. Rapid identification of *Campylobacter* spp. by melting peak analysis of biprobes in real-time PCR. J. Clin. Microbiol. 39: 2227-2232.

Oste, C. 1988. Polymerase chain reaction. BioTechniques. 6: 162-167.

Rasmussen, R. 1992. Optimising rapid cycle DNA amplification reactions. The RapidCyclist. Idaho Technology 1: 77.

Ririe, K.M., Rasmussen R.P. and Wittwer, C.T. 1997. Product differentiation by analysis of DNA melting curves during the polymerase chain reaction. Anal. Biochem. 245: 154-160.

Roche Molecular Biochemicals. Optimisation Strategy. Technical Note No. LC 9/2000.

Teo, I.A., Choi, J.W., Morlese, J., Taylor G. and Shaunak S. 2002. LightCycler qPCR optimistion for low copy number target DNA. J. Immuno. Methods. 270: 119-133.

Wittwer, C.T., Herrmann, M.G., Moss A.A. and Rasmussen, R.P. 1997. Continuous fluorescence monitoring of rapid cycle DNA amplification. BioTechniques. 22: 130-138.

5

Internal and External Controls for Reagent Validation

M. A. Lee, D. L. Leslie and D. J. Squirrell

Abstract

PCR applications that require a high confidence in the result should be designed to control for the occurrence of false negatives. False negatives can occur from inhibition of one or more of the reaction components by a range of factors. While an external, or batch control is often used, the ideal control is one that is included in the reaction cocktail in a multiplex format. Early approaches used different sized amplicons combined with end-point analysis. Fluorescent homogenous real-time PCR methods have a number of advantages for implementing internal controls. Here we discuss the application and development of molecular mimics for use as controls in real-time PCR, and explain a number of concepts and experimental considerations that will aid in the optimisation of the controlled multiplexed assay.

Introduction

The confidence in assays based on the polymerase chain reaction (PCR) (Saiki *et al.*, 1985; Mullis *et al.*, 1987) may be compromised by the sporadic occurrence of either false positive or false negative results. False positives are a problem that is common to the general application of PCR. False positives occur mainly as a result of cross contamination from either positive samples or reaction products. A combination of preventative methods including good laboratory practice, delineated preparation/analysis areas, PCR cabinets with UV treatment, UV air scrubbers, closed tube assays, and uracil glycosylase carry-over prevention chemistry (Longo *et al.*, 1990), has effectively eliminated the occurrence of false positives for the majority of applications. However, there are a number of PCR applications where the avoidance of false negative results is of equal importance. False negatives occur through failure of one or more of the reagents, the presence of inhibitors, or the failure of the PCR thermal cycling process (Rossen *et al.*, 1992; Wilson *et al.*, 1997). The applications requiring high confidence in the PCR include pathogen detection in clinical diagnosis, food quality control and environmental analysis.

For most molecular tests the use of reference material in a batch test is the only control that can be implemented. However, PCR and other nucleic acid amplification techniques provide not only extremely sensitive detection, but also have the added advantage that they can be readily multiplexed to include an internal control (IC). An IC is a second target molecule that can be amplified with, but distinguished from, other products in the same tube. In an ideal assay this should be able to control for all of the reagents in a reaction cocktail, and for variations in machine operation parameters. This can be achieved by the use of a molecular mimic, a synthetic molecule that may be co-amplified using the same set of amplimers as the target species.

Early approaches for ICs used amplicons with different molecular masses and subsequent analysis using a separation method. For example, the molecular mimic could contain both priming sites, but an internal sequence changed by insertions or deletions (Ursi *et al.*, 1992). The products of amplification could be easily analysed by the use of agarose gel electrophoresis and ethidium bromide staining. There are a large number of such examples reported in the literature. The main drawback of this approach is that the analysis method is time consuming, the smallest amplicon tends to be amplified more efficiently, and generation of specific and control product heteroduplexes (Henley *et al.*, 1996) can be unpredictable and therefore confusing for data interpretation.

Real-time PCR approaches using fluorescence are particularly useful for high confidence applications because they can be easily multiplexed. The assays are also generally carried out in a closed tube format which significantly reduces the risk of cross-contamination by amplification products. Most commercial fluorimeter instruments can co-amplify and detect at least two targets in a quantitative manner as discussed further in Chapter 3. However, increasing the level of multiplexing adds significantly to the complexity of the design strategy for the controls and the assay optimisation.

In this short chapter we will present various strategies for producing internal controls for different fluorescent chemistries, and discuss some experimental considerations for their optimisation and implementation. In the literature the term internal control is often used incorrectly and so we will first discuss some definitions. The experimental considerations described here are based largely on our own experiences and we make some useful suggestions for the development and implementation of controlled assay systems. Whilst users will find that for many applications the use of a homologous or competitive control is not necessary, the information will be of equal use to those developing heterologous controls and synthetic standards alike. Competitive homologous internal controls are state-of-the-art for PCR and the main focus of this chapter.

Nomenclature

In the literature multiplex PCR is used for a number of applications and the second internal (in the same reaction tube) amplicon usually serves to control, normalise or standardise the result. This chapter covers the application of molecular mimics to check reaction outcomes. However, the strategies described are equally applicable in the development of other internal and external mimics. In particular the use of mimics as either external or internal standards has advantages and in Table 1 we summarise some nomenclature used to describe the types of second amplicon. It is easy to see how a homologous internal control will add more confidence to a result. However, it should be noted that for controls that utilise nucleic acid reporter probes it is almost impossible to control for their function. This is discussed in more detail in the next section.

Strategies for Development of Internal Controls and their Analysis

The choice of strategy for detecting an internal control is dependent upon the instrument's capabilities and the reporting chemistry. The simplest strategy is one that uses a strand-specific probe using a reporting dye with a different emission wavelength to that of the probe for the specific target. This requires that the emission wavelength of these dyes is either spectrally separated enough that they do not overlap, or that these emissions can be effectively deconvoluted using the instrument's calibration algorithm. For example on the ABI 7700, a 5' nuclease assay may use FAM as the specific reporter and either VIC, JOE, HEX etc as the reporter for the control probe. A cycle-by-cycle increase in the emission of either dye reports the amplification of the respective species. This strategy may be applied to virtually all fluorescent chemistries and instruments.

Since the internal control serves only to report the integrity of the reagents the signal does not necessarily need to be quantitative. The ability to detect the species at the end of thermal cycling should suffice most applications. Therefore, for those assays based on hybridisation, and when an instrument that can determine melting point (T_m) is used, one can design the control amplicon and/or probe to have a different T_m from that of the specific nucleic acid target. This allows discrimination between target and control species and may be achieved by using DNA binding dyes for amplicons with different melting points (Lee

Table 1. Definitions and types of standard, control and reference nucleic acids.

Type \ Definition	Standard	Second amplicon type — Control	Second amplicon type — Reference
Definition	A standard is a nucleic acid preparation that has a set or known concentration. Unknown samples may be compared to one or more standards and an absolute value for the unknown can be determined by numerical interpolation/extrapolation from experimentally determined values.	A control is an amplicon added to the PCR and allowed to amplify to verify the integrity of one or more reagent(s) in the cocktail	A nucleic acid target that is used to compare the sample for either a qualitative or relative quantitative analysis.
External: Amplified in a different well.	The standards are amplified separately from the target species in different wells on the same instrument. *	"Batch control" This involves the use of a control in a separate well on the instrument This approach cannot control for false negatives in different samples. It controls only for successful thermal cycling and the integrity of a common core reaction mix.*	A related nucleic acid that is used for a qualitative comparison of amplification.
Internal: Amplified in the same well.	*Homologous* A competitive standard that has the same amplimers as the specific target. *Heterologous* A non-competitive standard with different amplimers to those of the specific target	*Homologous* A competitive control that is included in the same well of the instrument that has the same amplimers as the specific assay and controls for the core reagents and primers, and the thermal cycling. *Heterologous* A control that is included in the same well of the instrument but has different amplimers to that of the specific assay that controls for both the thermal cycling and the core reagents.	**Exogenous** This is an internal reference that is added to the sample and is co-amplified with the target nucleic acid using a second set of amplimers. **Endogenous** A "Housekeeping gene." This is usually a gene that is naturally present in the sample at constant levels such that the up or down regulation of the target species can be relatively quantified using a second set of amplimers.

et al., 1999), or labelled probes for different sequences. When the melting point method is used with sequence specific probes each may be associated with a dye that has a different emission wavelength. The merits of strand-specific probes over DNA binding dyes have been discussed in chapter 3. Whilst DNA binding dyes have some disadvantages in terms of specificity, the same probe is used for detecting the control and the specific target. Therefore, their function may be somewhat better controlled than strand-specific probes. Using dual hybridisation probes in either quantification and/or melting point analysis it is possible to use one common probe, either the donor or the acceptor, this also provides a better approach for control of the probe. The methods for generating these mimics are now briefly described.

Synthesising and Optimising Molecular Mimics for Use as Internal Controls

To generate a homologous internal control mimic there are several different approaches using recombinant DNA techniques. In each case the objective is to create a synthetic target that will effectively co-amplify with the specific amplicon, and whose "motif" may be easily distinguished from that of the specific amplicon. This is achieved by the use of dyes with different emission wavelengths and/or melting point analysis as described above. Several methods have been reported. These include: insertion of the specific target/amplimers into a vector and insertion/deletion of a sequence between the amplimers (Ursi *et al.*, 1992; Zimmerman *et al.*, 1996; Brightwell *et al.*, 1998), the amplification of a generic target from a related sequence (deWit *et al.*, 1993), the use of overlapping and tailed PCRs (Müller *et al.*, 1998; Sachadyn *et al.*, 1998), site directed mutagenesis (Nash *et al.*, 1995), and polymerase extension of oligonucleotides and subsequent PCR (Rosenstraus *et al.*, 1998). In addition a complete synthesis of the gene is possible for smaller amplicons (Zimmerman *et al.*, 2000) and some commercial suppliers will now make the gene and supply it in a vector ready for use. Since the ability to discriminate amplicons by size is not a requirement, the amplicon may be similar in size, and thus achieve a similar amplification efficiency. The two main techniques that we use to generate mimics are:

- Direct insertion of amplimer and probe sequences into cloning vectors using recombinant techniques. This can be done by amplifying amplimer/probe sites, or the use of oligonucleotide linkers. In either case the use of restriction overhangs will assist in the correct directional insertion of the fragments.

- Generation of targets using a PCR approach of amplimers tailed with specific amplimer sequences to amplify alternate probe regions. The products contain the specific amplimer sequences incorporated into the PCR product. This product may be subsequently cloned into the desired vector (See Figure 1). The use of the additional overhangs with unique restriction sites will assist in directional cloning that may be required for some applications (*e.g.* sense-specific RT-PCR).

For reverse transcriptase PCR (RT-PCR) a mimic may be made using the methods described and subsequent insertion of the product into a vector. *In-vitro* transcription using the T7/T3 promoters allows the RNA molecule to be then synthesised in quantity. The sequence could be cloned into an RNA virus such as the bacteriophage MS2. Armoured RNA™ is a product available from Ambion Inc (Pasloske, *et al.,* 1998). This is a method for packaging recombinant RNA into pseudoviral particles using a plasmid system whereby the MS2 bacteriophage coat protein is located downstream of the promoter and the target gene of interest. Under the control of the inducer, the RNA is transcribed encoding the coat protein and the target sequence. Coat protein is translated and subsequently binds to recombinant RNA to form the product. The packaged molecule is resistant to RNAase digestion yet easily extracted for amplification. The process has been shown to stabilise the RNA compared to *in vitro* transcripts made from plasmid and is an ideal method for RNA IC production.

Using any approach there are a number of experimental considerations that should be considered in the design and use of the control molecule and these will now be discussed.

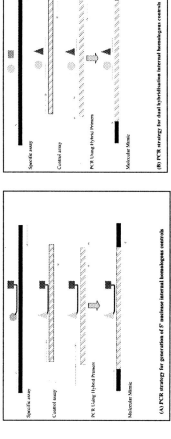

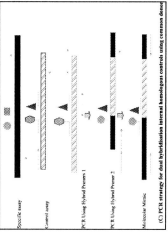

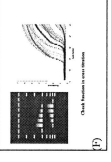

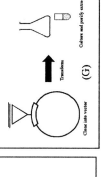

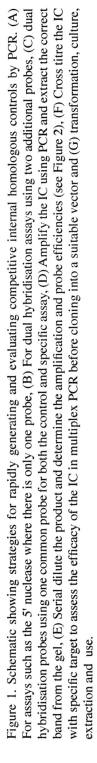

Figure 1. Schematic showing strategies for rapidly generating and evaluating competitive internal homologous controls by PCR. (A) For assays such as the 5' nuclease where there is only one probe, (B) For dual hybridisation assays using one common probe for both the control and specific assay, (C) dual hybridisation probes using two additional probes, (D) Amplify the IC using PCR and extract the correct band from the gel, (E) Serial dilute the product and determine the amplification and probe efficiencies (see Figure 2), (F) Cross titre the IC with specific target to assess the efficacy of the IC in multiplex PCR before cloning into a suitable vector and (G) transformation, culture, extraction and use.

Experimental Considerations

In designing an experiment the sequence of the specific target and the assay are important considerations. Reference should be made to chapter 3 covering the utility, design and sequence dependency of different chemistries. The design of the assay is best facilitated by software tools that can automatically select amplimers and probes and predict (score) amplimer-amplimer and amplimer-probe interactions for the multiplexed assay. Here we highlight some of the considerations that are important in achieving a successful IC implementation.

Control Sequence

The choice of control sequence is the most important factor since the other considerations are mostly dependent on this. When using nucleic acid probes one useful approach is to use a sequence that has been previously reported/optimised, or that is commercially available for the same assay type. For example, the ABI Human β-actin sequence for a 5' nuclease assay is useful because the probe is efficient and a kit containing it is commercially available. Ideally the probe sequence should not be of the same origin (or endogenous to the sample) as the target for the specific test. For example, it would not be ideal to use the human β-actin probe for a human test. Providing the probe sequence has been shown to work, a sequence from almost any other source could be suitable. Ideally, the GC content for the sequence should be similar to that of the specific target.

Amplicon Efficiency

Whilst the differences in amplification efficiency may be compensated for by changing the concentration of the initial target they should ideally be similar and this can be empirically determined. This is most important if the mimic is to be used for competitive quantification since the amplification efficiency can vary significantly between amplicons (Zimmerman *et al.*, 1996). It is useful to include native sequence on either side of the amplimer when designing the mimic to ensure that the effects of neighbouring sequence is maintained. The amplification efficiency is best determined using a DNA binding dye such as SYBR® Green I combined with an efficient hot-start to amplify a broad range or concentrations across the dynamic range of the assay. Whilst using a probe directly to do this is satisfactory, checking the amplification efficiency without the probe allows a direct measurement of efficiency and is useful in understanding the efficiency of the probe when it is later evaluated. It will avoid dismissing a satisfactory reaction when it may only be necessary to have a poor probe batch re-synthesised or to revise the probe design. The efficiency (E) can be calculated by using the slope value for the derived standard curve using (Figure 2):

$$E = 10^{-1/slope}$$

The efficiency of different amplicons for the same (or similar) dilution series can then be directly determined.

Probe (Nucleic Acid) Efficiency

Probe efficiency is important if the signal of the control is to be commensurate with that of the specific probe. The probe efficiency can be affected by a number of factors that are discussed in chapter 3. The probe sequence to be chosen should be checked for complementarities

Figure 2. Determining and comparing amplification and probe efficiencies of different amplicons. A: Determining amplification efficiency (E). Ideally, using a generic method such as a DNA binding dye, *e.g.* SYBR®Green-1, the cycle threshold (CT) values are determined for a dilution series of target material over a range of concentrations and the plot of fluorescence intensity against cycle number generates the amplification curves. The derived CT values can be used to generate a standard curve. The slope of the curve can then be used to calculate E for the range tested. This value can be used to compare E with other amplicons using the same or similar dilution series. B: Determining probe efficiency. Using the 5' nuclease assay and other assays by comparing the total change in fluorescence (as a function of a reference probe/assay) when the reaction is allowed to go to completion. C: Using probes based on hybridisation by comparing (as a function of a reference probe/assay) the change in fluorescence when the probe is hybridised (at a specific temperature) to when it is un-hybridised (at high temperatures).

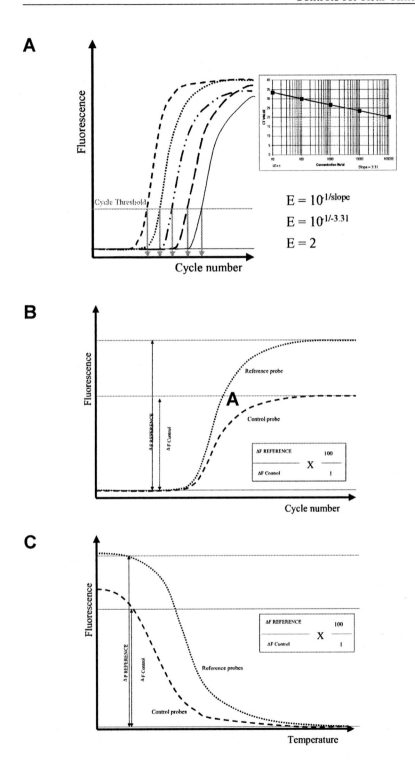

$$E = 10^{1/slope}$$
$$E = 10^{1/-3.31}$$
$$E = 2$$

to other probes and amplimers in the reaction. This is most important when the probe melting point approach is used since the signal from a control probe that is too large may swamp a low signal obtained in positives. When assessing potential probes for the mimic sequence it is useful to score the probe efficiency relative to a reference assay. For example, the percentage of signal:noise for both 5' nuclease assay specific and control probes can be compared to that of another assay such as the ABI Human β-actin probe. For this to be done the reaction for either probe must be taken to completion (plateau). This data may be expressed as:

$$\frac{\text{Fluorescence signal value specific assay/control assay at (cycle 50- cycle 1)}}{\text{Fluorescence signal value } \beta\text{-actin (Cycle 50-cycle 1)}} \quad X \quad \frac{100}{1}$$

This provides a comparative means for matching candidate control sequences to given assays. Likewise, the same may be done for comparing dual hybridisation probes using:

$$\frac{\text{Fluorescence signal value specific assay/control assay at (95°C- 60°C*)}}{\text{Fluorescence signal value } \beta\text{-globin (95°C- 60°C*)}} \quad X \quad \frac{100}{1}$$

*or at the temperature where the fluorescence data are to be acquired.

This approach may be carried out using any reference probe and is equally applicable to other assay systems to estimate the relative probe efficiency (see Figure 2). The quotient used to compare should be based on the values derived from the maximum signal (when the target is in excess of the probe) subtracted from the minimum signal (the background) for each probe. The minimum signal is usually at the start of the amplification or when the probe is disassociated from the target at elevated temperatures.

Control Vector

The choice of control vector is important since the vector may affect the stability and amplification efficiency. Plasmid constructs are commonly used and the efficiency of amplification may drop dramatically at lower target numbers (<100 copies per reaction). Linearising the plasmid by digestion at a unique restriction site in a vector can improve amplification efficiency. The use of other vectors such as bacteriophage λ vectors and M13 (ssDNA rescue by co-infection with helper phage) may also be useful approaches for generating large amounts of linear IC molecules. PCR product itself is fine for use as a control. We recommend the initial amplification of control DNA by PCR since the products may be used to evaluate potential mimics prior to the cloning of the molecule into a vector. However, the use of an *in vivo* amplification method for the final synthesis of DNA that is to be employed as the control is superior in that individual clones may be readily sequenced, to check their correct identity, and this process avoids the spurious artefacts generated in PCR. These artefacts can be minimised during the evaluation stages by cutting the predicted PCR band out of an agarose gel and using an appropriate method, such as column purification and/or dialysis, to clean up the product before it is diluted and evaluated as a suitable mimic.

Number of Molecules

When using a competitive control the number of molecules to be added into the reaction must be empirically determined so that the required sensitivity of the specific assay is not compromised. The control must be amplified effectively in the absence of specific target. With increasing numbers of specific target molecules in the reaction the control amplification may be out-competed. We have observed that it is possible to prevent this if the Tm of the specific amplicon is higher than that of the control. Reducing the denaturing temperature of the thermal cycle (which is possible on some thermal cyclers) to below that of the specific target, but higher than that of the control, suppresses specific amplification thus allowing the IC to amplify. However, this is not necessary since the amplification of the control only serves to validate the reagents in the absence of the specific target. A cross titration of specific target to control should be carried out to determine the correct amount of control molecules to include in a reaction. In practice, this amount is routinely determined to be 10-100 copies per reaction if the specific assay is to maintain a sensitivity of

10 or more copies of specific target. Accurate determination of the IC mimic concentration is therefore important. The use of fluorescent dyes and standards is inherently more accurate than ultra-violet optical density quantification at 260 nM (UV OD_{260}). An alternative method for statistically determining the correct concentration of a stock of internal controls using PCR is provided by Rosenstaus *et al.*, 1998. This method describes using a dilution series of the control preparation and performing multiple amplifications. At a given dilution the amplification will become stochastic due to the presence of low or no copies of the template in the reaction in accordance with Poisson's law. The relationship between the average number of internal control copies in a dilution (C) and the probability (Pn) that no molecules exist in a sample of this dilution is given by:

$$C=-ln(Pn)$$

Where, Pn is determined by counting the number of negative replicates used to calculate C.

Target Nucleic Acid

The control must be RNA if it is to be used for controlling the RT step. If a specific primer initiated RT step is used then utilising the same primers in the mimics will control for both the RT and the PCR, and negates the requirement of a second mimic for the PCR step in a one-step PCR. Depending on the application, controls at both the RNA and DNA levels may be required in a two-step PCR. The RT efficiency of the mimic and the specific target must be similar if the mimic is to be used as a standard in any application. When the RT-PCR is designed to be sense-specific, directional cloning and the choice of vector orientation is important.

Dye/Channel Assignment

The most efficient dye and/or detector should be used for the specific target assay. This is of importance where a single excitation source is used in a universal acceptor arrangement, in which case it is usually the shorter wavelength dye that will be most efficiently excited. When a universal donor arrangement is used, it will be the acceptor to which energy is most efficiently transferred from the donor. In either multiplex arrangement, the optimum channel will be dependent on the instruments specification and the dyes used.

Summary and Future Improvements

The further development of instruments and methods of detection will increase the ability to multiplex PCR to higher levels in a quantitative manner. This introduces new possibilities for controlling for/and or normalising sample extraction, and/or RT-PCR steps. This will be achieved through a combination of competitive (homologous) and non-competitive (heterologous) internal controls for each process. It is not always possible to control for these processes with one control since each may be subjected to different inhibitory factors. The ability to do this will improve the usefulness of the PCR for both research and diagnostic applications. The ability to control for all such processes will be of most use in emerging applications where remote testing in the non-laboratory environment necessitates the use of internal validation.

Acknowledgements

The authors would like to thank the editors for the invitation to contribute to this text. The underlying work leading to this publication was funded by the UK Ministry of Defence.

References

Brightwell, G., Pearce, M., and Leslie, D. 1998. Development of internal controls for PCR detection of *Bacillus anthracis*. Mol. Cell. Probes. 12: 367-377.

deWitt, D., Wooton, M., Allan, B., and Steyn, L. 1993. Simple method for production of internal control DNA for *Mycobacterium tuberculosis* polymerase chain reaction assays. J. Clin. Micro. 31: 2204-2207.

Henley, W.N., Schuebel, K.E., and Nielsen, D.A. 1996. Limitations imposed by heteroduplex formation on quantitative RT-PCR. Biochem. Biophys. Res. Comm. 226: 113-117.

Lee, M.A., Brightwell, G, Bird, H., Leslie, D., and Hamilton, A. 1999. Fluorescent detection techniques for real-time multiplex strand specific detection of Bacillus anthracis using rapid PCR. J. Applied. Micro. 87: 218-223.

Longo, M.L., Berninger, M.S., and Harley, J.L. 1990. The use of uracil glycosylase to control carry-over contamination in polymerase chain reactions. Gene 93: 125-128

Müller, F-M.C., Schnitzler, N., Coot, O., Kockelkorn, P., Haase, G., and Li, Z. 1998. The rationale and method for constructing internal control DNA used in Pertussis polymerase chain reaction. Diag. Microbiol. Infect. Dis. 31: 517-523.

Mullis, K.B., and Faloona, F.A. 1987. Specific synthesis of DNA *in vitro* via a polymerase catalyzed chain reaction. Meth. Enzymol. 155: 335-351.

Nash, K.A, Klein, J.S., and Inderlied, C.B. 1995. Internal controls as performance monitors and quantitative standards in the detection by polymerase chain reaction of herpes simplex virus and cytomegalovirus in clinical specimens. Mol. Cell. Probes. 9: 347-356.

Pasloske, B.L., Walkerpeach, C.R., Obermoeller, R.D., Winkler, M., and DuBois, D.B. 1998. Armoured RNA technology for production of ribonuclease-resistant viral RNA controls and standards. J. Clin. Micro. 36: 3590-3594.

Rosenstraus, M., Wang, Z., Chang S-Y, DeBonville, D., and Spadoro, J.P. 1998. An internal control for routine diagnostic PCR: Design properties, and effect on Clinical performance. J. Clin. Micro. 36: 191-197.

Rossen, L., Norskov, P., Holmstrom, K., and Rasmussen, O.F. 1992.

Inhibition of PCR by components of food samples, microbial diagnostic assays and DNA extraction solutions. International J. Food. Micro. 17: 37-45.

Sachadyn, P. and Kur, J. 1998. The construction and use of a PCR internal control. Mol. Cell. Probes. 12: 259-262.

Saiki, R.K., Scharf, S., Faloona, F.A., Mullis, K.B., Horn, G.T., Erlich, H.A., and Arnheim, N. 1985. Enzymatic amplification of beta-globin genomic sequences and restriction site analysis for diagnosis of sickle cell anemia. Science. 230: 1350-1354.

Ursi, J-P., Ursi, D., Ieven, M., and Pattyn, S.R. 1992. Utility of an internal control for the polymerase chain reaction. Application to detection of *Mycoplasma pneumoniae* in clinical specimens. APMIS 100: 635-639.

Wilson, I.G. 1997. Inhibition and facilitation of nucleic acid amplification. Appl. Environ. Microbiol. 63, 3741-51.

Zimmerman, K, and Mannhalter, J.W. 1996. Technical aspects of quantitative competitive PCR. BioTechniques. 21: 268-279.

Zimmerman, K., Rieger, M., Groß, P., Turecek, P.L., and Schwarz, H.P. 2000. Sensitive single-stage PCR using custom synthesised internal controls. BioTechniques. 28: 694-702.

6

Quantitative Real-Time PCR

N. A. Saunders

Abstract

Unlike classical end-point analysis PCR, real-time PCR provides the data required for quantification of the target nucleic acid. The results can be expressed in absolute terms by reference to external quantified standards or in relative terms compared to another target sequence present within the sample. Absolute quantification requires that the efficiency of the amplification reaction is the same in all samples and in the external quantified standards. Consequently, it is important that the efficiency of the PCR does not vary greatly due to minor differences between samples. Careful optimisation of the PCR conditions is therefore required. The use of probes in quantitative real-time PCR improves its performance and a range of suitable systems is now available. Generally quantitative real-time assays have excellent performance characteristics including a wide dynamic range, high sensitivity and accuracy. This has led to their use in a wide range of applications and two examples are presented. Viral quantification is now an important factor in the control of infection. The problems associated with virus quantification in cytomegalovirus (HCMV) infection are similar to those presented by other viruses. Quantitative PCR is finding an increasing role in the diagnosis of cancer. The assessment of *c-erbB2/Her2/neu* gene duplication is useful in predicting the disease prognosis in breast cancer. Several different real-time quantitative PCR protocols are available for these applications and have been applied successfully to their respective diseases.

Introduction

The early years of the PCR were not without difficulty. One of the greatest barriers to successful exploitation was the problem of contamination that becomes evident due to the high sensitivity of the method. It was also quickly realised that the term 'quantitative PCR' could almost be considered an oxymoron due to the non-linearity of the relationship between amplicon yield and starting copy number. The combined effect of these factors was that even a single amplicon contaminating a reaction mixture could give a false positive result. The practical uses of PCR were therefore restricted and there remained many applications requiring highly sensitive and quantitative nucleic acid detection assays that could not be addressed. This led to the development of fixes that rendered PCR at least semi-quantitative. This was generally achieved by limiting the number of temperature cycles to ensure that reactions containing template in the desired quantitative range did not pass the linear phase of product accumulation. Truly quantitative but relatively inconvenient methods were also developed. These were based on either the competitive target or limiting dilution approaches. These methods are used for RNA and DNA copy number measurement in applications such as viral load estimation and gene expression studies. The widespread use of quantitative PCR was therefore already well established by the time real-time PCR machines were introduced.

The problem with using PCR quantitatively stems from the variable kinetics of product accumulation after different numbers of cycles. Although the process is essentially exponential during the early cycles this rate cannot be maintained when nanogram/μl quantities of product are present. The main reasons for this are the increasing level of amplicon reannealing during the priming step and the overstretching of reaction resources including polymerase, primers and nucleotides as the quantity of DNA to be synthesized during the elongation phase increases. During this phase product accumulation is approximately linear for a number of cycles. Finally, if cycling is continued beyond the linear phase the quantity of amplicon may reach a plateau when new synthesis of full-length product is only sufficient to replace material lost by the combined effects of hydrolysis at elevated temperatures and exonuclease activity.

There are now two main approaches used to work around the problematic kinetics of PCR in order to obtain accurate quantitative data. The first is competitive quantitative PCR that relies on the assumption that a competitor target added to the PCR reaction will be amplified at the same rate as the authentic target provided that they have identical primer binding sites and broadly similar structures. The input level of target molecules is calculated from the final ratio of the two products. The main disadvantage of this method is that its precision depends upon the accuracy with which the target to competitor ratio can be determined at the end of the amplification process. Generally, this ratio can be measured with greatest accuracy when it is close to unity. This limits the range of quantitative values that can be determined accurately in a single tube containing a set number of copies of the competitor molecule. Running parallel tubes containing different numbers of copies of the competitor can extend the accurate quantitative range of competitive PCR.

The second main approach to quantification by PCR relies on the use of real-time machines. The rationale behind the method depends upon the experimental evidence and theoretical prediction that product accumulation within PCR reactions generally remains exponential for many cycles. While the concentration of product remains very low the effects of sample variation on the amplification efficiencies of different reactions remain small. For example, any low-level polymerase inhibitors introduced with the samples are less likely to prevent the completion of the full-length second strands when the polymerase is present in large excess. The exponential phase only ends when either the level of product reaches a level of several $ng/\mu l$ or when the reaction is 'poisoned' due to the accumulation of primer artefacts. This means that during the exponential phase there is a linear relationship between $\log_{10}$ starting template copies and the number of

A

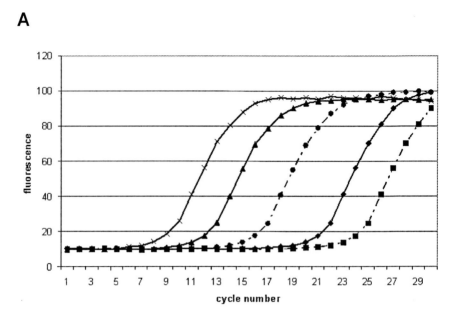

B

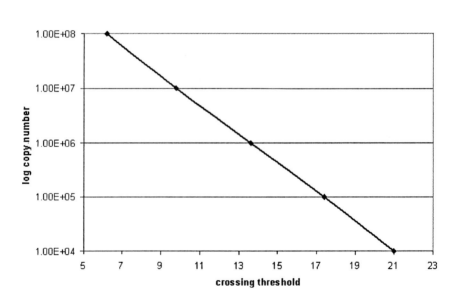

Figure 1. Quantitative real-time PCR. Panel A represents amplification profiles for a series of $\log_{10}$ dilutions of a standard target at 1×10^4 (squares) to 1×10^8 (crosses) copies per reaction. Panel B shows the straight line relationship between the copy number and crossing threshold.

amplification cycles required for the product to reach a set or threshold concentration (Figure 1). In theory the threshold should be set at the lowest reliably measurable level, since the accuracy of quantification should be highest when the measurements are made at the earliest possible stage within the exponential phase. This general approach has proved to be very accurate and reliable.

The measurement of product threshold levels in real-time quantitative PCR is done by analysis of the plots of fluorescence against cycle number. The simplest approach is to set a threshold fluorescence value and then determine the number of cycles required for the fluorescence to reach this level. The threshold level can be set automatically by the software which determines the standard deviation of the points forming the baseline. The threshold is then set at, for example, three standard deviations above the baseline mean. An alternative method is to determine the second derivative maximum for each curve describing fluorescence versus cycle and use this as the threshold value (LightCycler® Online Resource Site at http://www.lightcycler-online.com/). This is the point at which the rate of change of fluorescence is fastest and usually occurs at the end of the exponential phase. The fluorescent signal is relatively high at this point and can therefore be measured relatively accurately. The second derivative maximum method looks promising in theory but generally in practice the accuracy of quantitative PCR is greater when the data are analysed by the baseline method.

Standards and Controls for Quantitative Real-Time PCR

Inhibition Controls

PCR inhibitors introduced into the reaction, usually with the sample, may present a barrier to successful quantification. Although it can be assumed that low levels of inhibition will have no effect on the efficiency of PCR before the crossing threshold has been reached, this will not be the case at higher inhibitor levels. In some situations it may be adequate to identify any samples containing inhibitors from their 'flat' curves or from the complete absence of an amplification product. However, it may be necessary to include experimental controls that indicate the presence of inhibitory materials (Figure 2). The choice is between internal and external control templates. External controls are simply PCR targets that are amplified under the same conditions as the test material. These are spiked into parallel reaction tubes that are otherwise identical to the experimental tubes. Although it would be possible to use the test sequence as the external control this approach is not advised as it may cause or complicate the investigation of instances of PCR contamination. A better solution is to use a modified version of the target sequence that has been modified by deletion, insertion or substitution within the inter-primer region so that it can be identified specifically in real-time (Kearns *et al.*, 2001; Kearns *et al.*, 2002). The advantage of using an external control is that it cannot affect the conditions within the experimental reaction vessel. The main disadvantage is that an additional reaction is needed for each sample. Internal controls follow the same design as external controls but are spiked into the experimental reaction mixture. Following amplification

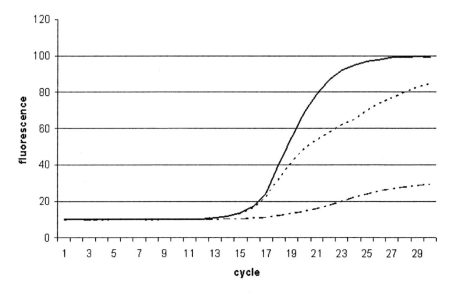

Figure 2. Effect of inhibitors on amplification profiles. The solid line shows an uninhibited reaction in comparison with a reaction that contains some inhibitory material (dotted) and one that is more strongly inhibited (dots and dashes).

it is necessary to distinguish between the two products and this can be achieved by using detection probes that can be distinguished in the real-time instrument. It is also essential to ensure that the internal control is added at the lowest concentration that is consistently detectable. This reduces the possibility of interference between the two reactions that might decrease the efficiency of experimental target amplification. One consequence of the use of low levels of the internal inhibition control is that, due to competition, it will not be amplified to the detection threshold if the target sequence is present at high concentrations. However, under such circumstances the inhibition control is redundant.

External Quantification Controls

Conventionally external quantification controls have precisely the same sequence as the target to be quantified. However, it may be advantageous to introduce a minor sequence modification to allow amplicons derived from the control sequence to be distinguished from the target amplicons in cases of suspected contamination. If the modified base is placed at the probe site it should be possible to identify the quantification control in real-time. Sequence variations at other sites must be detected by alternative means. Stocks of reliably quantified standards of the desired sequence can be prepared using the protocol outlined below.

The accuracy of quantification by real-time PCR depends on the quality of the standard curve. The copy number range of the standards should always encompass the range expected in the experimental samples. The maximum increment between quantitative standards should be $\log_{10}$ and a minimum of four points should be included in each curve. The question arises of whether it is necessary to prepare a separate standard curve for each machine run. Whilst this is undoubtedly the best practice and should give the greatest quantitative accuracy since it eliminates inter-run variation it may not always be a practical solution. An intermediate solution would be to prepare standard curves periodically but to include one quantified standard in each run to check for any drift.

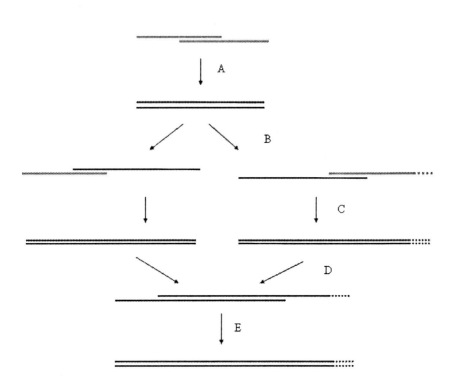

Figure 3. Synthesis of standards for use in real-time quantification assays. Synthetic oligonucleotides are shown in grey and products in black. In step A, primers with complementary 3' ends are mixed and subjected to PCR cycling in the presence of thermostable polymerase and deoxynucleotide substrates. This reaction produces a double stranded product that is mixed with two further primers (step B) complementary to the new 3' ends. If RNA standards are required a T7 RNA polymerase promoter sequence is added to the 5' end of one primer (dotted line). In steps C to E PCR cycling is again used to produce a double stranded product. Step C is the addition of the sequences carried by the second set of primers to the complementary strands. Step D is the annealing of the products of step C which are converted into the double stranded form in step E. Amplification of the final product then continues under the direction of the second primer pair. It may be possible to mix all of the primers into a single mixture but this may not give a good yield in all cases and purification of the first product may be required.

A Simple Approach for the Preparation of Inhibition and Quantification Controls for Real-Time Quantitative PCR

The first step is to design the control template *in silico*. For inhibition controls this is done simply by modifying the target sequence at its probe binding site. It is best to rearrange the relevant bases so that the level of homology between the old and new probe-binding sites is <50%. The new sequence should be identical to the original sequence in terms of the primers and amplification conditions required but should not be detected by the original probe. For quantification controls a single base substitution is sufficient to allow the control to be identified. Check that no stable secondary structures haves been introduced into the sequence using suitable computer software (*e.g.* OLIGO 6; Molecular Biology Insights, Cascade, Colorado).

If the inhibition control is to be approximately 140 bases or less, it is simplest to produce it by PCR synthesis starting with extended primers (up to 70-80 bases) that have 20 base complementary ends. Longer controls can be built up in steps of up to 120 bases using additional extended primers (Figure 3). Following the final stage of amplification the control template should be gel purified (*e.g.* QIAquick Gel Extraction kit; Qiagen, Crawley, West Sussex) and then carefully quantified. Alternatively, the control sequence may be cloned into a plasmid vector (*e.g.* using the TOPO kit, Invitrogen, Paisley, Scotland), grown in *Escherichia coli* cells and then purified using a suitable mini-prep kit. It is advisable to verify the sequence of the cloned insert at this point. The simplest and in many cases the most reliable method of quantification is to compare the intensity of a stained band with a commercial quantitative gel standard (*e.g.* 100bp Ladder; New England Biolabs, Hitchin, Herts). This method is most accurate when the bands compared have approximately equal intensities.

Stocks of standard materials should be stored frozen at the highest available concentration in Tris EDTA (TE) buffer. Dilutions of the standards should be made in a solution of carrier nucleic acid (*e.g.* 5μg/ml of sheared salmon sperm DNA dissolved in deionised water) to prevent the loss of significant numbers of copies adhering to the walls of the storage vessel. It is also important to minimise the loss of the standard DNA by avoiding repeated freeze/thaw cycles.

If RNA standards are required a T7 RNA polymerase promoter sequence should be included at the 5' end of one of the primers (Figure 3). The PCR product may then be used as the substrate in a reaction containing T7 RNA polymerase and the four nucleotides. Residual template can be removed from the reaction using RNase-free DNase prior to quantification of the RNA. RNA standards should be handled and stored very carefully to avoid introducing RNase. One option for reducing the potential problem of nuclease damage is to add an RNase inhibitor such as RNasin (Promega, Madison, WI) to all stocks.

Probes

Quantitative real-time PCR is possible when double stranded DNA binding dyes (*e.g.* SYBR Green I) are used to report the accumulation of the amplicon. However, the accuracy and reliability of this method depends upon the quality of the reaction chemistry to a greater extent than when a sequence specific probe detection method is used. The reason for this is that in the dye system PCR artefacts can be interpreted as copies of the target molecule. When oligonucleotide probe reporters are used this problem does not arise.

Although not recommended for quantitative PCR, the DNA binding dye methods have the potential advantage that the level of signal generated is relatively high and this results in high signal:noise ratios. High signal to noise ratios give smooth plots of fluorescence against cycle number and allow crossing thresholds or maximum delta fluorescence (dF) values to be determined with great precision. Since the signal:noise ratio of the fluorescent signal is important in determining the accuracy of quantification by real-time PCR it must also be an important consideration in choosing the probe system to be employed. It is important to select a system that generates strong signals (see Chapter 3) since this will minimise the variation between replicate measurements.

Dynamic Range, Sensitivity and Reliability of Real-Time Quantification

Dynamic Range

A characteristic feature of real-time PCR is that it offers accurate quantification over a very wide range of concentrations. At the top end of the range the limit is set by the need to establish a base-line level of fluorescence before product accumulation is detected, since after this point it is not possible to determine an accurate crossing threshold. In practice this means that it is possible to measure up to approximately 10^{10} copies of template. The bottom end of the quantification scale is set by the sensitivity of the PCR. This may be increased by pre-amplification of the target sequence using external primers (*i.e.* nested PCR). Care must be taken to ensure that the pre-amplification reaction does not progress beyond the exponential phase. This can be done by using a limited number of first-round cycles (Brechtbuehl *et al.*, 2001). Although nested reactions can detect single copies of the target sequence it is unrealistic to expect accurate quantification at very low copy numbers due to sampling errors.

Linearity of Real-Time Quantitative PCR

Plots of $\log_{10}$ initial template copies against the number of cycles required to reach the crossing threshold (or second derivative maximum) in real-time PCR are expected to yield a straight line. This relationship may break down at low copy numbers resulting in either an increase or a decrease in the number of cycles required to give a $\log_{10}$ increase in copy number. Apparent increases in the efficiency of the reaction may occur at low copy number when the intercalating method of product detection is used. This is due to detection of the accumulated double stranded primer artefacts after many PCR cycles. Real decreases in PCR efficiency actually occur as primer artefacts reach high levels due to the depletion of reaction resources. This decrease becomes evident when sequence-specific detection methods are used.

Accuracy

There are two main sources of error in quantification assays that rely upon real-time PCR. The first is the accuracy with which the reaction threshold can be measured. This is largely determined by the level of noise in the baseline signal and depends on the signal to noise ratio. Different real-time machines and detection chemistries vary in this parameter. Probably the most important factor is that the signal transduction chemistry chosen gives a strong signal. The second major source of error is likely to be variations in the sample that have small but cumulatively significant effects on the amplification efficiency. A single-cycle difference between measurements of the crossing threshold will result in a two-fold under- or over-estimation of the initial copy number. Errors of this magnitude are difficult to eliminate completely and in some systems may be even higher. In contrast, sample pipetting errors have a relatively small effect on the overall accuracy of real-time quantification. Although these should be minimised they are unlikely to contribute significantly to the overall accuracy of the real-time quantification.

The errors in real-time PCR quantifications can be minimised and measured by performing replicate analyses. In general, as noted for pipetting errors, variations during sample preparation are relatively small compared with those introduced in the real-time PCRs. Consequently, although ideally many replicates of multiple samples should be tested, in practice it is most advantageous to test a single sample preparation in many real-time tubes. As a minimum, testing in triplicate is recommended.

Reaction Efficiency

The efficiency of PCR reactions can be estimated from the slope of the crossing threshold against copy number line. Most PCR reactions involving short products with normal levels of secondary structure are close to 100% efficient so that a $\log_{10}$ increase in the concentration of the target amplicons requires approximately 3.3 cycles. Efficiency is equal to $E=10^{-1/slope}$ so that 100% efficiency gives a value of 2

(Bernard and Wittwer, 2002). Higher apparent levels of efficiency are difficult to explain from PCR theory and may indicate that the standards have not been prepared accurately or that other errors have occurred. Poor optimisation of the PCR reaction conditions can result in low efficiency. For example, annealing at too high a temperature or denaturing at a temperature that is too low will both reduce the PCR efficiency. Often the low efficiency of a PCR reaction results from secondary structures within the template. An example of this would be any structure at one of the primer binding sites that inhibits annealing. Low efficiency PCRs (*i.e.* E=1.5) can be used for quantification without compromising accuracy. Long PCR products are also associated with poor reaction efficiency. The main reason for this, at least during the early rounds of the PCR, is misincorporation of bases resulting in chain termination which occurs at a higher frequency per round of amplification for longer products. The major disadvantage of the use of an inefficient PCR is that the dynamic range is unlikely to extend to the lower copy numbers.

Inhibitor Effects

PCR inhibitors are substances that either reduce the activity of the polymerase directly or that compete with the target template limiting the availability of polymerase or other resources. At the start of the PCR the polymerase, primers and nucleotides are all present in great excess and consequently any inhibitors may have zero net effect on the efficiency of the reaction. The effect of any inhibition will then increase roughly in proportion with the quantity of DNA to be synthesised. Accurate quantification is not possible for samples that contain inhibitors that reduce the efficiency of amplification significantly before the reaction threshold has been crossed. It is therefore important to use extraction methods that eliminate all non-trivial sources of PCR inhibition.

Absolute or Relative Quantification?

For absolute quantification the number of target template copies present within a unit quantity of sample is determined. This depends upon the reliability of target nucleic acid extraction and may therefore present a problem for cellular material that is difficult to lyse. It also depends, crucially, on the quality and uniformity of the sample. So, for example, sputum samples examined for the presence of pathogenic organisms are likely to present serious problems for absolute quantification due to their non-uniformity. One effective solution to these difficulties is to perform relative quantifications in which the measured quantity of target sequence is expressed as a ratio with the quantity of a reference sequence as the denominator. In the example of tissue samples a suitable reference might be a human gene. Levels of induced/repressed mRNAs can be expressed most usefully as comparisons with quantities of housekeeping gene transcripts or rRNA subunits. The estimation of the levels of gene duplication in cancer cells is another application where ratios are far more meaningful than absolute quantification values (Bernard and Wittwer, 2002).

Examples of the Use of Quantitative Real-Time PCR

The Measurement of Human Cytomegalovirus (HCMV) Levels in Infected Human Serum

Primary infection with HCMV is usually asymptomatic although occasionally it may lead to a mononucleosis-like syndrome with prolonged fever, and a mild hepatitis. Although the immune system suppresses the virus in healthy individuals, infection results in lifelong persistence of the viral genome. Congenital infection is more significant and results in serious complications in a proportion of cases. Recurrent disease may occur if the immune system is suppressed and this is associated with morbidity and mortality especially in bone marrow or solid-organ transplant recipients and in patients with AIDS. Early diagnosis of active infection is important in high-risk patients since antiviral (ganciclovir) treatment is available and pre-emptive therapy reduces the incidence and severity of disease. Consequently, patients at risk are usually monitored by methods such as pp65 antigen detection or viral DNA PCR (Boeckh and Boivin, 1998). Quantitative PCR for HCMV has been greatly simplified by the use of real-time machines and is becoming the method of choice for early detection of rises in the HCMV titre or for monitoring the progress of antiviral therapy (Kearns *et al.*, 2001).

Several real-time quantitative PCR assays for HCMV have been reported in the literature (Gault *et al.*, 2001; Guiver *et al.*, 2001; Kearns *et al.*, 2001; Cortez *et al.*, 2003). Most of these methods provide absolute quantification of number of copies of viral DNA per millilitre of serum. However, in one case (Sanchez and Storch, 2002) the quantity of the viral DNA was normalised to the quantity of human DNA present. The sequences targeted in these studies varied although the conserved gene encoding glycoprotein B was a popular choice. In general, the length of the PCR products was rather longer than might be preferred for real-time PCR. It has been reported that serum HCMV is highly fragmented (Boom *et al.*, 2002) and consequently the use of long PCR targets might be expected to impose an additional limit on the sensitivity of these assays. The quantified external controls used in these assays were produced by cloning the PCR product into a suitable plasmid vector. These control plasmids are generally linearised and then quantified. In one study (Kearns *et al.*, 2001) the cloned sequence was modified at two bases within the probe binding site. This has the advantage that the PCR product of plasmid amplification can be distinguished from viral product by the probe melting temperature and thus contamination of samples with plasmid DNA may be recognised. Internal controls that are amplified by the HCMV specific primers but with a different internal sequence were used in the protocol described by Cortez (Cortez *et al.*, 2003). The internal control, which is added to all mixes as an inhibition check, is distinguished from the HCMV product by using different hybridisation probe pairs. Protocols that use either hybridisation (double check) or hydrolysis (TaqMan) probes on either a block based (ABI 7700) or air (LightCycler) have been described. The relative merits of these instruments are discussed elsewhere (Chapter 2).

Assessment of *c-erbB2/Her2/neu* Gene Duplication in Breast Cancer Tissue

An increase in the number of copies of cellular oncogenes caused by DNA amplification often leads to deregulated protein expression and is associated with neoplastic transformation. It has been reported (Slamon *et al.*, 1987) that *c-erbB2/Her2/neu* gene amplification from 2 to >20 fold is found in 30% of human breast tumours. The degree of amplification is correlated with a poor prognosis for a range of tumours including those of breast tissue (Slamon *et al.*, 1987; Press *et al.*, 1997). Investigation of the significance of gene duplication requires that the analytic methods are reliable, accurate and sensitive. Accuracy is required because the extent of duplication is relatively limited. High sensitivity is needed because tissue biopsies are often of limited weight and it may be necessary to obtain material by microdissection (Glockner *et al.*, 2000; Lehmann *et al.*, 2000).

The development of quantitative real-time PCR methods has greatly facilitated the measurement of gene duplication in breast cancer (Glockner *et al.*, 2000; Lehmann *et al.*, 2000; Bernard and Wittwer, 2002; Gunnarsson *et al.*, 2003; Konigshoff *et al.*, 2003). The published assays generally use either the LightCycler or the ABI 7700 combined with appropriate signal transduction chemistries. In contrast with the methods for quantification of HCMV, which usually provide an absolute measure of the number of copies of the virus, methods for measuring *c-erbB2/Her2/neu* gene duplication usually provide a ratio. Several different denominator genes have been used including those for albumin, β-globin, the amyloid precursor and deoxycytidine kinase. These are all single copy genes in normal cells. One group (Glockner *et al.*, 2000; Lehmann *et al.*, 2000) employed two control genes. This approach has the advantage that it is more likely to reveal cases of control gene duplication that might cause underestimation of the experimental gene copy number. All of the methods cited above measure relative gene copy numbers by analysis of the experimental and control targets within a single tube. If all of the PCR reactions in the multiplex have identical levels of efficiency, the amplification curves of the different targets will have the same crossing threshold, providing no gene duplication has occurred. In tumours with gene duplication of one of the targets the crossing threshold for the duplicated gene shifts to a lower value. This method is able to detect gene duplication with great sensitivity. The magnitude of the duplication can then be estimated accurately from the crossing threshold shift and the efficiency of the PCR.

One group (Lyon et al., 2001; Millson et al., 2003) have devised an interesting method that uses the capabilities of a real-time PCR instrument (the LightCycler) to perform quantitative competitive PCR. The competitor was a modified version of the genomic c-erbB2/Her2/neu target that differed by a single base substitution in the probe binding site. The target and competitor could be effectively distinguished by the 8°C difference between the melting curve maxima. Plots of the ratio of competitor to target melting peak area against competitor copy number for a fixed amount of test genomic DNA gave a good fit to a third-order polynomial curve. Over a limited range of ratios the linear fit was also reasonably accurate. For practical detection of c-erbB2/Her2/neu gene duplication, 50ng aliquots of test DNA were tested against 15,000 copies of competitor sequence. This single point was used to find the copy number from the known third-order polynomial curve. The gene dose estimations by this method for twenty tumour samples correlated well with values determined by quantitative real-time PCR.

Conclusions

Real-time quantitative PCR is now well established as a method for estimating quantities of specific nucleic acid sequences. One major advantage is that it may be applied to samples containing levels of the target sequence that are several $\log_{10}$ units lower than can be detected using hybridisation based methods. Another is that this sensitivity contributes to a dynamic range that can be >8 $\log_{10}$ units in a single assay. In comparison with alternative methods such as competitive PCR real-time quantification is simpler and more accurate. Relative quantification is particularly well suited to the new real-time instruments since the control and experimental genes may be tested under identical conditions in the same tube.

The drawbacks of quantitative real-time PCR are few and minor, with the exception of the capital outlay needed for the initial purchase of a real-time instrument. However, the cost of the instrumentation is declining rapidly in real terms and this is likely to continue in parallel with increases in performance, reliability and ease of use. The cost per test of quantitative real-time PCR (excluding capital cost) is similar to that of qualitative real-time PCR which can now be lower overall than that of standard PCR with an agarose gel end-point analysis.

Future Developments

Real-time PCR instrument development continues to deliver machines that are more rapid, accurate and sensitive than there forerunners. In addition, the associated control and analysis software is becoming more intuitive and capable. In parallel with the advances in the instrumentation, improved chemistries are delivering greater analytical sensitivity and improved specificity. These improvements are resulting in an increase in the accuracy of quantitative PCR based on relative crossing threshold measurements.

Arrays are now being used for quantitative assays, particularly for measurement of the levels of individual mRNAs within complex mixtures. Clearly, such parallel systems are extremely attractive in the appropriate circumstances. The main limitations of arrays are their relatively low sensitivities and dynamic ranges. In future, improved instrumentation and systems for amplification on the chip may become more viable and allow arrays to encroach further into the domain of quantitative PCR. However, for the present real-time PCR generally provides higher quality data than can be obtained from arrays and should remain an important complementary method due to the reliability of the quantitative data generated.

References

Bernard, P. S., and Wittwer, C. T. 2002. Real-time PCR technology for cancer diagnostics. Clin. Chem. 48: 1178-1185.

Boeckh, M., and Boivin, G. 1998. Quantitation of cytomegalovirus: methodologic aspects and clinical applications. Clin. Microbiol. Rev. 11: 533-554.

Boom, R., Sol, C. J., Schuurman, T., Van Breda, A., Weel, J. F., Beld, M., Ten Berge, I. J., Wertheim-Van Dillen, P. M., and De Jong, M. D. 2002. Human cytomegalovirus DNA in plasma and serum specimens of renal transplant recipients is highly fragmented. J. Clin. Microbiol. 40: 4105-4113.

Brechtbuehl, K., Whalley, S. A., Dusheiko, G. M., and Saunders, N. A. 2001. A rapid real-time quantitative polymerase chain reaction for hepatitis B virus. J. Virol. Methods 93: 105-113.

Cortez, K. J., Fischer, S. H., Fahle, G. A., Calhoun, L. B., Childs, R. W., Barrett, A. J., and Bennett, J. E. 2003. Clinical trial of quantitative real-time polymerase chain reaction for detection of cytomegalovirus in peripheral blood of allogeneic hematopoietic stem-cell transplant recipients. J. Infect. Dis. 188: 967-972.

Gault, E., Michel, Y., Dehee, A., Belabani, C., Nicolas, J. C., and Garbarg-Chenon, A. 2001. Quantification of human cytomegalovirus DNA by real-time PCR. J. Clin. Microbiol. 39: 772-775.

Glockner, S., Lehmann, U., Wilke, N., Kleeberger, W., Langer, F., and Kreipe, H. 2000. Detection of gene amplification in intraductal and infiltrating breast cancer by laser-assisted microdissection and quantitative real-time PCR. Pathobiology 68: 173-179.

Guiver, M., Fox, A. J., Mutton, K., Mogulkoc, N., and Egan, J. 2001. Evaluation of CMV viral load using TaqMan CMV quantitative PCR and comparison with CMV antigenemia in heart and lung transplant recipients. Transplantation 71: 1609-1615.

Gunnarsson, C., Ahnstrom, M., Kirschner, K., Olsson, B., Nordenskjold, B., Rutqvist, L. E., Skoog, L., and Stal, O. 2003. Amplification of HSD17B1 and ERBB2 in primary breast cancer. Oncogene 22: 34-40.

Kearns, A. M., Guiver, M., James, V., and King, J. 2001. Development and evaluation of a real-time quantitative PCR for the detection of human cytomegalovirus. J. Virol. Methods 95: 121-131.

Kearns, A. M., Turner, A. J., Eltringham, G. J., and Freeman, R. 2002. Rapid detection and quantification of CMV DNA in urine using LightCycler-based real-time PCR. J. Clin. Virol. 24: 131-134.

Konigshoff, M., Wilhelm, J., Bohle, R. M., Pingoud, A., and Hahn, M. 2003. HER-2/neu gene copy number quantified by real-time PCR: comparison of gene amplification, heterozygosity, and immunohistochemical status in breast cancer tissue. Clin. Chem. 49: 219-229.

Lehmann, U., Glockner, S., Kleeberger, W., von Wasielewski, H. F., and Kreipe, H. 2000. Detection of gene amplification in archival breast cancer specimens by laser-assisted microdissection and quantitative real-time polymerase chain reaction. Am. J. Pathol. 156: 1855-1864.

Lyon, E., Millson, A., Lowery, M. C., Woods, R., and Wittwer, C. T. 2001. Quantification of HER2/neu gene amplification by competitive PCR using fluorescent melting curve analysis. Clin. Chem. 47: 844-851.

Millson, A., Suli, A., Hartung, L., Kunitake, S., Bennett, A., Nordberg, M. C., Hanna, W., Wittwer, C. T., Seth, A., and Lyon, E. 2003. Comparison of two quantitative polymerase chain reaction methods for detecting HER2/neu amplification. J. Mol. Diagn. 5: 184-190.

Press, M. F., Bernstein, L., Thomas, P. A., Meisner, L. F., Zhou, J. Y., Ma, Y., Hung, G., Robinson, R. A., Harris, C., El-Naggar, A., Slamon, D. J., Phillips, R. N., Ross, J. S., Wolman, S. R., and Flom, K. J. 1997. HER-2/neu gene amplification characterized by fluorescence *in situ* hybridization: poor prognosis in node-negative breast carcinomas. J. Clin. Oncol. 15: 2894-2904.

Sanchez, J. L., and Storch, G. A. 2002. Multiplex, quantitative, real-time PCR assay for cytomegalovirus and human DNA. J. Clin. Microbiol. 40: 2381-2386.

Slamon, D. J., Clark, G. M., Wong, S. G., Levin, W. J., Ullrich, A., and McGuire, W. L. 1987. Human breast cancer: correlation of relapse and survival with amplification of the HER-2/neu oncogene. Science 235: 177-182.

7

Analysis of mRNA Expression by Real-Time PCR

Stephen A. Bustin and Tania Nolan

Abstract

The last few years have seen the transformation of the fluorescence-based real-time reverse transcription polymerase chain reaction (RT-PCR) from an experimental tool into a mainstream scientific technology. Assays are simple to perform, capable of high throughput, and combine high sensitivity with exquisite specificity. The technology is evolving rapidly with the introduction of new enzymes, chemistries and instrumentation and has become the "Gold Standard" for a huge range of applications in basic research, molecular medicine, and biotechnology. Nevertheless, there are considerable pitfalls associated with this technique and successful quantification of mRNA levels depends on a clear understanding of the practical problems and careful experimental design, application and validation.

Introduction

The conventional reverse transcription polymerase chain reaction (RT-PCR) (Simpson *et al.*, 1988; Vrieling *et al.*, 1988) is unrivalled as a sensitive assay for the rapid, inexpensive and simple detection of RNA. Importantly, the ability to amplify several templates in a single reaction (multiplexing) (Edwards *et al.*, 1994) has made it a truly high throughput assay (Willey *et al.*, 1998).

However, for a long time quantitative RT-PCR was perceived as a qualitative assay capable of answering yes/no questions only, since even minor variations in reaction components and thermal cycling conditions can greatly affect the yield of any amplified product (Wu *et al.*, 1991). However, the publication of several ground-breaking reports addressing the fundamental theoretical and practical issues associated with quantification (Kelley *et al.*, 1993; Nedelman *et al.*, 1992; Raeymaekers, 1993; Siebert *et al.*, 1992) led to a more general recognition of the inherent quantitative capacity of the PCR assay (Halford *et al.*, 1999). This development was advanced by the obvious need for quantitative data, *e.g.* for measuring viral load in HIV patients (Kappes *et al.*, 1995), monitoring of occult disease in cancer (Bustin *et al.*, 1998) or examining the genetic basis for individual variation in response to therapeutics through pharmacogenomics (Jung *et al.*, 2000).

Conventional quantitative protocols rely either on competitive techniques where known amounts of RT-PCR-amplifiable competitors are spiked into the RNA samples prior to the RT step (Freeman *et al.*, 1999) or on non-competitive methods where the target is co-amplified with a second RNA molecule with which it shares neither the primer recognition sites nor any internal sequence (Reischl *et al.*, 1999). However, the lack of standardisation produces strikingly inconsistent results (Apfalter *et al.*, 2001; Mahony *et al.*, 2000; Smieja *et al.*, 2001), with variable reproducibility (Henley *et al.*, 1996; Souaze *et al.*, 1996), false positive rates as high as 28% (Johnson *et al.*, 1999) and error rates ranging between 10% and 60% dependent on the analysis method (Souaze *et al.*, 1996; Zhang *et al.*, 1997a; Zhang *et al.*, 1997b). Equally as important, conventional quantification protocols are, without exception, labour- and reagent-intensive (Raeymaekers, 1993), with post-PCR processing particularly likely to result in subjective interpretation of results.

The real-time fluorescence-based quantitative RT-PCR (qRT-PCR) combines the amplification and analysis steps of the PCR reaction, thereby eliminating the need for post-PCR processing. It does this by monitoring the amount of DNA produced during each PCR cycle and its sensitivity, specificity and wide dynamic range have revolutionised the approach to PCR-based quantification of RNA, making it the method of choice for quantitating steady-state mRNA levels (Bustin, 2000). The method is based on the principle that there is a quantitative

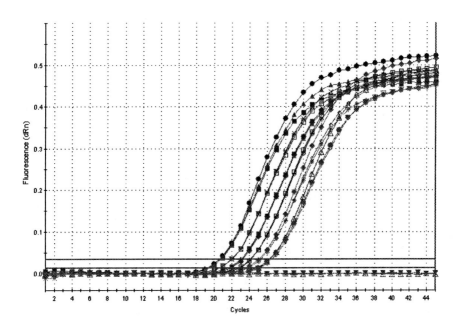

Figure 1. Determination of CT. First, the levels of background fluorescence are established for a particular run. Next, platform-specific algorithms are used to define a fluorescence background threshold. Finally, the algorithm searches the data from each sample for a point that exceeds the baseline. The cycle at which this point occurs is defined as the CT. It is dependent on the starting template copy number, the efficiency of PCR amplification, efficiency of cleavage or hybridization of the fluorogenic probe and the sensitivity of detection. The fewer cycles it takes to reach a detectable level of fluorescence, the greater the initial copy number.

relationship between the amount of target RNA present at the start of the assay and the amount of product amplified during its exponential phase. The crucial conceptual innovation and the key to understanding quantification by qRT-PCR is the threshold cycle (CT) (Figure 1).

The development of robust chemistries (as discussed in chapter 3) and the cost-reduction caused by the introduction of several competing combined thermal cycler/fluorimeters has converted qRT-PCR into a simple and relatively affordable assay and has transformed it from an experimental tool into a mainstream scientific technology (Ginzinger, 2002). Its convenience (Ke *et al.*, 2000a), detection range (Gerard *et al.*, 1998), low inter-assay variation (Wall S.J. *et al.*, 2002) and reproducibility (Max *et al.*, 2001) have made it the method of choice for the quantification of mRNA expression (Bustin, 2000; Bustin, 2002). Currently its only real drawback is the substantial monetary investment in instrumentation and reagents that is required to carry out an assay.

RNA Purification

A cell's ability to adjust its mRNA copy numbers in response to environmental changes is a crucial element in the complex regulation of gene expression. In addition, a wide spectrum of pathological processes is associated with changes in gene expression at the mRNA level. These alterations are regulated by numerous endogenous and exogenous stimuli and are the result of at least two processes: (1) increased or reduced transcription caused by highly complex combinatorial control mechanisms (Hill *et al.*, 1995) and (2) changes to mRNA stability, with the degradation rates of different mRNAs varying over a broad range (Ross, 1995).

Therefore, extraction and analysis of cellular RNA provides a snapshot of steady-state mRNA levels and the preparation of the RNA is arguably the most important determinant of a meaningful qRT-PCR experiment. This is tricky as (1) the original expression profile must be conserved, (2) RNA must be of the highest quality, (3) it should be free of DNA and (4) no inhibitors of the RT or PCR steps must be co-purified. The first problem is that the sources of mRNA are disparate and include not just undemanding *in vitro* material such as tissue culture cells, but *in vivo* biopsies obtained during endoscopy, post-surgery, post-mortem and from archival materials. Naked RNA is extremely susceptible to degradation by endogenous ribonucleases (RNases) that are present in all living cells. Therefore, the key to the successful isolation of high quality RNA and to the reliable and meaningful comparison of qRT-PCR data is to ensure that neither endogenous nor exogenous RNases are introduced during the extraction procedure (Lee *et al.*, 1997). A second problem for experiments quantitating mRNA levels is that mRNA expression profiles can change rapidly after cells or tissue samples have been collected, but before they have been frozen. This is a particular problem with cells extracted from whole blood (Keilholz *et al.*, 1998) and mRNA profiles can change over several orders of magnitude even in the short time it can take from collecting the blood to processing it in the laboratory. Causes are RNA degradation, the induction of certain genes, *e.g.* Cox-2, as well as the method of RNA preparation (de Vries *et al.*, 2000). Furthermore, blood is notorious for containing numerous inhibitors of the PCR reaction (Akane *et al.*, 1994; Al Soud *et al.*, 2001; Fredricks *et al.*, 1998). Therefore, appropriate sample acquisition has a major influence on the quality of the RNA and subsequently on any result of qRT-PCR assays (Vlems *et al.*, 2002) and once a biological sample has been obtained, immediate stabilization of RNA is the most important consideration (Madejon *et al.*, 2000).

Recent reports suggest that RNA extracted from formalin-fixed, paraffin-embedded archival materials can be successfully quantitated by real-time RT-PCR assays. Between 60 % (Mizuno *et al.*, 1998) and 84% (100% of samples less than 10 years old) (Coombs *et al.*, 1999) of templates can be amplified by RT-PCR, and it is possible to quantitate mRNA expression levels (Cohen *et al.*, 2002; Godfrey *et al.*, 2000; Specht *et al.*, 2001; Stanta *et al.*, 1998a; Stanta *et al.*, 1998b), even after immunohistochemical staining (Fink *et al.*, 2000). This is exiting, since retrospective analysis of archival tissue could enable the correlation of molecular findings with the patient's response to treatment and eventual clinical outcome. Real-time RT-PCR assays, with their amplicon lengths of around 100 bp, are ideally placed to amplify the usually degraded RNA from these archival samples whose average size is 200-250 nucleotides (Bock *et al.*, 2001; Lehmann *et al.*, 2001).

Just as real-time chemistries have revolutionized the PCR reaction, so the development of the Agilent Bioanalyzer together with the RNA 6000 LabChip has revolutionized the ability to analyze RNA samples for quality. This technology is without doubt the method of choice for analyzing RNA preparations destined to be used in qRT-PCR assays. The Agilent 2100 Bioanalyzer is a small bench top system that integrates sample detection, quantification, and quality assessment. It does this by using a combination of microfluidics, capillary electrophoresis and fluorescent dye that binds to nucleic acid. Size and mass information is generated by the fluorescence of the nucleic acid molecules as they migrate through the channels of the chip. The instrument software quantitates unknown RNA samples by comparing their peak areas to the combined area of the six reference RNAs. It has a wide dynamic range and can quantitate as little as 5 ng/µl, although it is most accurate at concentrations above 50 ng/µl. An electropherogram of fluorescence versus time and a virtual gel image are generated, and the software assesses RNA quality by using the areas under the 28S and 18S rRNA peaks to calculate their ratios. The analysis of a total RNA sample from two human colon biopsies is shown in Figure 2.

qRT-PCR Assay

A successful qRT-PCR experiment is made up of several steps that must be considered carefully before starting the experiment. Variability of any one can result in unreliable quantification and render the assay meaningless.

mRNA or Total RNA

First, a decision must be made whether to use mRNA or total RNA as the target for the RT-PCR assay. There is some suggestion that starting with mRNA results in better sensitivity (Burchill *et al.*, 1999) but any

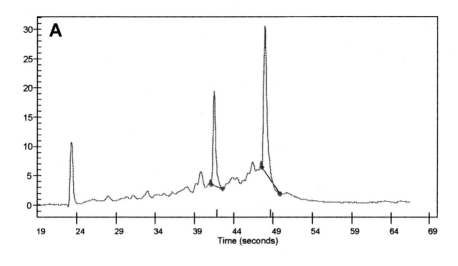

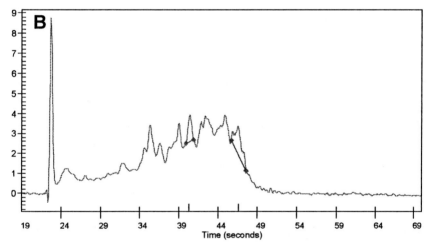

Figure 2. Electropherogram generated using the Agilent 2100 Bioanalyzer. A: An example of a high quality RNA preparation with distinct 28S and 18S peak ratios. The amount of small RNA (peak 1), which can include 5.8S and 5S ribosomal peaks and transfer RNA, is highly dependent on the preparation protocol. B: An example of a poor RNA preparation, with lots of additional peaks indicative of the presence of degraded RNA.

inhibitory effects of background RNA are noticeable only when very small numbers of target RNA template (<10) are present (Curry *et al.*, 2002). There are several reasons for preferring the use of total RNA: (1) purification of mRNA involves an additional step and any increased sensitivity may be cancelled by the possible loss of material, (2) not all mRNA molecules have polyA tails and (3) the concentration of mRNA may be insufficient to allow proper quality assessment. Therefore on balance, it is advisable to use total RNA, as the ultimate sensitivity of the qRT-PCR assay is likely to be determined by factors other than whether mRNA or total RNA was used in the first place.

Regardless of whether total RNA or mRNA is used, sample acquisition and purification of the RNA mark the initial step of every qRT-PCR assay. It cannot be over-emphasized that the quality of the template is arguably *the* most important determinant of the reproducibility and biological relevance of subsequent qRT-PCR results. Any problems that affect reproducibility are likely to have originated here (Bomjen *et al.*, 1996). Many samples, especially biopsies of human tissue, are unique, hence a wasted nucleic acid preparation means that the opportunity to record data from that sample is irretrievably lost. Consequently it is prudent to expend extensive efforts on getting every stage of this process absolutely right, starting with consistency when collecting, transporting and storing samples, continuing with rigorous adherence to extraction protocols and remaining vigilant when taking samples out of storage for analysis.

cDNA Priming

Priming of the cDNA reaction from the RNA template can be done using random primers, oligo-dT, or target-specific primers and the choice of primer can cause marked variation in calculated mRNA copy numbers (Zhang *et al.*, 1999).

Random Primers

This method is by definition, non-specific, but yields the most cDNA and is most useful for transcripts with significant secondary structure. However, since transcripts originate from multiple points along the transcript, more than one cDNA transcript is produced per original target. Furthermore, the majority of cDNA synthesise d from total RNA will be ribosomal RNA-derived and may compete with a target that is present at very low levels. The Tm of random primers and oligo-dT is low, hence neither can be used with thermostable RT enzymes without a low temperature pre-incubation step.

Oligo-dT

cDNA synthesis using oligo-dT is more specific than random priming, as it will not result in priming from rRNA. It is the best method to use when the aim is to obtain a faithful cDNA representation of the mRNA pool, although it will not prime any RNAs that lack a polyA tail. In addition, oligo-dT priming requires very high quality RNA that is full length, hence is not a good choice for priming from RNA that is likely to be fragmented, such as that obtained from formalin-fixed archival material. Furthermore, the RT may fail to reach the primer probe-binding site if secondary structures exist or if the primer/probe-binding site is at the extreme 5'-end of a long mRNA.

Target-Specific Primers

Target-specific primers synthesise the most specific cDNA and provide the most sensitive method of quantification (Lekanne Deprez et al., 2002). Their main disadvantage is that they require separate priming reactions for each target, which is wasteful if only limited amounts of RNA are available.

Choice of Enzyme

Viral RTs have a relatively high error rate, and have a strong tendency to produce truncated cDNA products because of pausing (Bebenek et al., 1989). Their lack of thermal stability is a particular problem as the significant secondary structure of many mRNA molecules requires their denaturation for efficient reverse transcription (Buell et al., 1978; Kotewicz et al., 1988; Shimomaye et al., 1989). Therefore, extensive efforts have been directed at improving these enzymes and finding new ones. Consequently, there are many enzymes to choose from (Table 1) and they have differing abilities to copy small amounts of template and transcribe RNA secondary structures accurately. For example, increasing the reaction temperature to 50°C (Freeman *et al.*, 1996) or 55°C (Malboeuf *et al.*, 2001), depending on the RT, enhances both the accuracy and the specificity of retroviral RTs as not only is the intra- and intermolecular base pairing reduced but false priming is also minimised.

Other strategies for dealing with RNA structure include supplementing the RT reaction with dimethyl sulphoxide (DMSO), which disrupts

Table 1. Reverse transcriptase enzymes.

Enzyme	Supplier	T_{opt} (°C)	RNaseH	Features
AMV-RT	Various	37	+++	
MMLV-RT	Various	45	+	
Omniscript	Qiagen	37	+	Not AMV- or MMLV-derived
Sensiscript	Qiagen	37	+	Not AMV- or MMLV-derived
Powerscript	Clontech	42	-	
Superscript II	Invitrogen	42-50	-	
StrataScript	Stratagene	42	-	Point mutation in RNaseH domain of MMLV-RT
ImProm-II	Promega	55	-	
MMLV PM	Promega	42-55	-	Point mutation in RNaseH domain of MMLV-RT
Tfl pol	Promega	60-72	-	one tube/one enzyme RT-PCR
Tth pol	Various	60-72	-	one tube/one enzyme RT-PCR
Expand RT	Roche	42	-	Point mutation in RNaseH domain of MMLV-RT
RevertAid	Fermentas	42-45	-	Point mutation in RNaseH domain of MMLV-RT
Thermoscript	Invitrogen	50-65	-	

base-pairing and can increase the yield during amplification. Sodium pyrophosphate (<4mM) may also be included in the RT step of two-step assays that use Avian myeloblastosis virus (AMV) RT to overcome RNA secondary structure and increases the yield of first-strand cDNA. The pyrophosphate slows the DNA polymerisation rate of the enzyme and is presumed to make it less susceptible to disassociating from the template at regions of strong secondary structure. However, this is not a recommended approach if using Moloney murine leukemia virus (MMLV)-RT since this enzyme has an increased sensitivity to pyrophosphate.

AMV RT is more robust and exhibits greater processivity than MMLV RT (Brooks *et al.*, 1995) and retains significant polymerisation activity up to 55°C (Freeman *et al.*, 1996), although it is usually used at 42°C in two-step RT-PCR. Its error rate is 4.9×10^{-4} based on misincorporation studies (Ricchetti *et al.*, 1990). Life Technologies offers the ThermoScript™ RT, a recombinant, genetically engineered AMV RNaseH^{-ve} RT with increased thermal stability compared with standard AMV-RT. This allows reactions to be run at temperatures up to 65°C to generate long cDNA transcripts. Native MMLV-RT has significantly less RNaseH activity than native AMV-RT (Gerard *et al.*, 1997), but is less thermostable and is typically used at 37°C. The error rate is slightly higher than that of AMV-RT (Ricchetti *et al.*, 1990) and several modified MMLV-RT-derived enzymes have been introduced that offer higher efficiency and improved synthesis of cDNA, *e.g.* PowerScript™ RT (Clontech), SuperScript™ II (Invitrogen) and THERMO-RT™ (Display Systems Biotech Inc).

In addition, there are several DNA-dependent DNA polymerases, *Tth* (Myers *et al.*, 1991), *Tfl* (Harrell *et al.*, 1994) and *Bca*BEST™ (Takara), that exhibit both RNA- and DNA-dependent polymerisation activities in the presence of Mn^{2+}. All three are genuinely thermostable enzymes. In practical terms, only the primer Tm limits the temperature at which the reverse transcription step can be carried out. The sensitivity of target detection using *Tth* polymerase can be extremely high, and the detection of low abundance mRNA from single cells has been reported (Chiocchia *et al.*, 1997). These enzymes do not have 3-5' exonuclease

135

activities, hence their error rates are comparatively high (but no higher than those of the viral RTs).

Amplification Stratagies

The RT-PCR can be performed as a one enzyme, one tube procedure (sometimes described as one-step) or a two-enzyme procedure, either as a one tube or a two-tube reaction assay. Single-tube RT-PCR procedures, whether one or two enzyme-based, are the techniques of choice in most clinical and high throughput laboratories because they are simple to use and offer a reduced risk of cross-contamination (Aatsinki *et al.*, 1994; Goblet *et al.*, 1989; Mallet *et al.*, 1995; Wang *et al.*, 1992). In two-tube RT-PCR, first-strand synthesis is performed in a small reaction volume, and an aliquot is then added to a PCR reaction for amplification of a specific target. However, the RT enzyme itself (Fehlmann *et al.*, 1993) and certain RNase inhibitors (Lau *et al.*, 1993) can inhibit the PCR step.

Two Enzyme Procedures: Separate RT and PCR Enzymes

In "uncoupled" two tube, two-enzyme procedures first-strand cDNA synthesis is performed in the first tube, under optimal conditions, using random, oligo-dT or sequence-specific primers. An aliquot of the RT reaction is then transferred to another tube, which contains the thermostable DNA polymerase, DNA polymerase buffer, and PCR primers, for PCR carried out under optimal conditions for the DNA polymerase. In "continuous" one tube, two-enzyme procedures the reverse transcriptase produces first-strand cDNA in the presence of Mg^{2+}, high concentrations of dNTPs, and either target-specific or oligo-dT primers. Following the RT reaction, an optimized PCR buffer (without Mg^{2+}), a thermostable DNA polymerase, and target-specific primers are added to the tube and PCR is performed. Alternatively, a single buffer can be used for both steps, obviating the need to open the tubes between the two steps. Interassay variation of two step RT-PCR assays can be very small when carried out properly, with correlation coefficients ranging between 0.974 and 0.988 (Vandesompele *et al.*,

2002a). The disadvantages of this approach are:

1. In two enzyme/one tube assays a template switching activity of viral RTs can generate artefacts during transcription (Mader *et al.*, 2001). This does not occur with bacterial polymerases.
2. Since in two enzyme/one tube assays all reagents are added to the reaction tube at the beginning of the reaction, it is not possible to separately optimize the two reactions.
3. Two enzyme/two tube reactions involve a significant amount of additional effort and present additional opportunities for contamination.

Single RT and PCR Enzyme Assays

A single enzyme able to function both as an RNA- and DNA-dependent DNA polymerase can be used in one tube/one enzyme procedures that are performed in a single tube without secondary additions to the reaction mix. However, there are several disadvantages to this approach:

1. Since all reagents are added to the reaction tube at the beginning of the reaction, it is not possible to separately optimize the two reactions.
2. The one tube/one enzyme assay is about 10-fold less sensitive than an uncoupled assay, due possibly to the less efficient RT activity of *Tth* polymerase (Cusi *et al.*, 1994; Easton *et al.*, 1994). However, it should be noted that one manufacturer (Promega) claims a one enzyme continuous assay is more sensitive than an uncoupled assay.
3. A direct comparison of one enzyme and two enzyme continuous assays has revealed more consistent sample-to-sample results with the latter.
4. This assay requires the use of significantly more units of DNA polymerase which can drive up the cost of the single enzyme assay.
5. The most thorough study comparing one step and two step reaction conditions using SYBR Green chemistry found that the one step

reaction was characterised by extensive accumulation of primer dimers, which could obscure the true results in quantitative assays (Vandesompele et al., 2002a).

Inhibitors

The co-purification of inhibitors of the RT-PCR reaction during template preparation can present a serious problem to accurate and reproducible quantification of mRNA levels (Cone et al., 1992). Common inhibitors include various components of body fluids and reagents encountered in clinical and forensic science (e.g., haemoglobin and urea), food constituents (e.g., organic and phenolic compounds and fats), and environmental compounds (e.g., humic acids and heavy metals) (Wilson, 1997). In addition, factors like DNA fragmentation (Golenberg et al., 1996;Pikaart et al., 1993) and the presence of residual anti-coagulant heparin (Beutler et al., 1990) or proteinase K-digested haem compounds like haemoglobin (Akane et al., 1994) or myoglobin (Belec et al., 1998) will negatively affect PCR efficiency. The problem with this type of inhibitor is that it makes the comparison of qPCR results from different patients, or different samples from the same patient impossible as it results in different amplification efficiencies and hence CTs of the same target from different patients. Worryingly, laboratory plastic ware has been identified as one potential source of PCR inhibitors (Chen et al., 1994). It is also important to remember that reagents can have a significant effect on assay reproducibility, with lot-to-lot variation an essential consideration (Burgos et al., 2002) and that DNA amplification products themselves can inhibit the polymerase (Kainz, 2000).

Different polymerases will react differently to inhibitors and it is worthwhile checking each template preparation for inhibition and testing several polymerases for their efficiency at amplifying the template. Sequence-dependent PCR amplification bias has been observed and results in incorrect or ambiguous quantification (Shanmugam et al., 1993) and can even result in the preferential amplification of one allele over another (Weissensteiner et al., 1996).

AmpliTaq Gold and the *Taq* polymerases are totally inhibited in the presence of 0.004% (v/v) blood, whereas HotTub (*Tfl*) from *Thermus flavus*, *Pwo*, r*Tth*, and *Tfl* DNA polymerases are able to amplify DNA in the presence of 20% (v/v) blood without reduced amplification sensitivity; the DNA polymerase from *Thermotoga maritima* (Ultma) appears to be the most susceptible to a wide range of PCR inhibitors. HotTub and *Tth* are the most resistant to the inhibitory effect of K^+ and Na^+ (Al Soud *et al.*, 1998) and biological samples (Poddar *et al.*, 1998; Wiedbrauk *et al.*, 1995). Thus, the PCR inhibiting effect of various components in biological samples can, to some extent, be eliminated by the use of the appropriate thermostable DNA polymerase. One thing to bear in mind is that some enzymes are in fact enzyme mixes that combine a highly processive enzyme with a proof-reading enzyme to balance fidelity and yield (Barnes, 1994). Finally, contamination of the PCR assay is an ever present danger (Kwok *et al.*, 1989).

Additives

Chemical additives have been reported as improving the reliability of "tricky" PCR assays (Pomp *et al.*, 1991). Just as there are numerous potential causes of any PCR problem, there are numerous reagents that are supposed to solve those problems. The list includes non-ionic detergents such as Tween 20 (Demeke *et al.*, 1992), additives such as glycerol and formamide (Comey *et al.*, 1991), the proteins bacteriophage T4 Gene 32 protein, gelatine and bovine serum albumin (Comey *et al.*, 1991, Kreader, 1996), tetramethylammonium chloride (TMAC) (Chevet *et al.*, 1995), the sugars sucrose and trehalose (Louwrier *et al.*, 2001) and even ethidium bromide (Hall *et al.*, 1995). These additives may stabilise the polymerase, enhance its processivity, disrupt template secondary structure and decrease melting temperature. There are commercially available additives such as Stratagene's Taq Extender™ PCR Additive and PerfectMatch® PCR Enhancer, CLONTECH's GC-Melt™ and Novagen's NovaTaq™ PCR Master mix. Three factors are important if additives are to be useful: high potency, high specificity and a wide effective range (Chakrabarti *et al.*, 2001), but, unfortunately, there is no consensus with respect to their effectiveness. Different additives have significantly different effects with different targets and, it requires empirical evidence, rather than theoretical deliberations, to choose the most appropriate one. In some amplifications, specificity is reported to be improved by formamide, but not DMSO (Sarkar *et al.*, 1990), whereas in others DMSO was more effective than formamide (Varadaraj *et al.*, 1994). Therefore, it is best to test a wide variety of potential additives and to choose the one that is the most effective in one's own particular assay.

The most common additive is DMSO (Winship, 1989), which is added at a concentration of 2-10 % and several other sulfoxides have been identified as novel enhancers of high GC target amplification (Chakrabarti *et al.*, 2002). DMSO decreases secondary structure, lowers the melting temperature of the primers, can suppress recombinant molecules forming during the PCR (Shammas *et al.*, 2001a) and works with *Tth* polymerase (Sidhu *et al.*, 1996). On the other hand, it inhibits *Taq* polymerase activity at concentrations higher than 10 %.

Trehalose may increase the enzymatic activity of MMLV-RT at 60°C (Carninci *et al.*, 1998) and the specificity of oligonucleotide-dT priming by Superscript II (Mizuno *et al.*, 1999), probably by increasing their thermal stability (Hottiger *et al.*, 1994). Another stabilising agent is betaine (N, N, N-trimethylglycine) (Henke *et al.*, 1997; Weissensteiner *et al.*, 1996), which is sometimes used in combination with DMSO (Baskaran *et al.*, 1996) or trehalose (Spiess *et al.*, 2002) and is present in a number of commercially available PCR enhancement kits. Betaine also increases the thermal stability of proteins (Santoro *et al.*, 1992), may prevent non-specific interaction of polymerase and template and reduce or even eliminate the base pair composition dependence of DNA thermal melting transitions by equalizing the contributions of GC and AT base pairs (Rees *et al.*, 1993). The contradictory effect obtained with such additives is underlined by a comparison of the amplification efficiency of GC rich templates in the presence of betaine with that of DMSO, glycerol, trehalose, or TMAC. The results showed that only betaine had any appreciable effect (McDowell *et al.*, 1998). Betaine also increases the tolerance of the RT-PCR assay to inhibition by heparin (Weissensteiner *et al.*, 1996) and reduces recombination events (Shammas *et al.*, 2001a; Shammas *et al.*, 2001b). Another problem with these additives is that the various reports recommend different optimal concentrations. This probably suggests that the effects are template and amplicon dependent and that careful titration and optimisation of their concentration prior to any assay is required for these additives to have any positive effects. This is especially so because adding too much has an inhibitory effect (Spiess *et al.*, 2002). Therefore, if the RT-PCR assay turns out to be a difficult one, it might be more productive to properly optimise the RT-PCR assay from the start, together with a possible re-design of primers and probes.

PCR Optimisation Strategies

Optimisation prior to carrying out the real-time PCR assay with precious samples will generate data that reflect on the quality of the assay design and produces valuable information to indicate where problems may lie. By using the minimum primer and probe concentration to give the best assay conditions it is often possible to reduce the concentration of oligonucleotides included in the assays and so increase the assay's specificity as well as saving costs. The simple steps detailed below are designed to be as labour, time and cost efficient as possible. Note that successful multiplex reactions depend critically on assay optimisation.

The optimisation process requires the use of a suitable template that specifies the target under investigation. This can be from any convenient source, and commonly used templates are a linear plasmid DNA containing the amplicon, a purified PCR fragment, genomic DNA (*only intron-less genes*) or cDNA. Optimisation of primers should be carried out using SYBR Green I. This allows the design of several alternative assays around the sequence of interest and the testing of several primer combinations before purchasing an internal oligonucleotide probe. Since SYBR Green I is a convenient and inexpensive approach, it provides reassurance that the primers are functioning and, with a melting curve, a measure of the quality of the reaction. Primer dimer products may be formed and will be visualized using melting profiles. Although primer dimer products are not detected using labelled internal oligonucleotide probes, substantial non-specific product formation could inhibit the PCR reaction and result in reduced efficiency. For this reason it is imperative to select primer pairs that produce little or no non-specific products in the PCR reaction.

Primer Optimisation

1. Dilute probe and primers to 100 μM for stock storage, aliquot and store at -70°C
2. Dilute primers to 10 μM for working solutions (store at -20°C)
3. Dilute probe to 5 μM for working solution (store at -20°C)
4. Select template DNA that can be used for optimisation

The optimal concentration of primers can span a surprisingly wide range from 50 nM to 900 nM and so a matrix of reactions encompassing this range is performed. The following optimisation reactions are run in duplicate:

1. Set up a matrix of 0.5 ml tubes as shown on Table 2. Each tube will eventually hold reaction mix to produce 2 x 25 μl duplicate optimisation reactions. These will then be divided into individual 25 μl reactions.
2. Make a reaction master mix by adding the reagents in the order shown in Table 3, mix gently and then finally add *Taq* polymerase and mix gently again. Add 38.4 μl of the reaction mix to each primer/water set-up tube in the matrix. Mix gently and then aliquot 25 μl of this into each of two instrument compatible microfuge tubes. Cap carefully, label tubes (if required), spin briefly, gently wipe the surface of each tube and ease into the PCR block noting the orientation of the tubes.
3. Perform a qPCR assay and melting analysis as described in Table 3. SYBR Green I data can be collected at either the extension stage or the latter half of the annealing stage.

To select the optimal primer pairs, examine the amplification plots and select the combination that give the lowest CT value. An example of a primer optimisation matrix is shown in Figure 3.

Table 2. Primer Optimisation Matrix set-up. R= Reverse (primer final concentration) F= Forward (primer final concentration). NOTE: all 50 nM final concentration primers use $1\mu M$ primer stocks, all other concentrations use $5\ \mu M$ primer stocks.

	μl		μl		μl		μl		μl
F (50nM) 1μM stock R (50nM) 1μM stock	3 3	F (50nM) 1μM stock R (100nM) 5μM stock	3 1.2	F (50nM) 1μM stock R (300nM) 5μM stock	3 3.6	F (50nM) 1μM stock R (600nM) 5μM stock	3 7.2	F (50nM) 1μM stock R (900nM) 5μM stock	3 10.8
ddH₂O	15.6	ddH₂O	21.4	ddH₂O	15	ddH₂O	11.4	ddH₂O	7.8
F (100nM) 5μM stock R (50nM) 1μM stock	1.2 3	F (100nM) 5μM stock R (100nM) 5μM stock	1.2 1.2	F (100nM) 5μM stock R (300nM) 5μM stock	1.2 3.6	F (100nM) 5μM stock R (600nM) 5μM stock	1.2 7.2	F (100nM) 5μM stock R (900nM) 5μM stock	1.2 10.8
ddH₂O	17.4	ddH₂O	19.2	ddH₂O	16.8	ddH₂O	13.2	ddH₂O	9.6
F (300nM) 5μM stock R (50nM) 1μM stock	3.6 3	F (300nM) 5μM stock R (100nM) 5μM stock	3.6 1.2	F (300nM) 5μM stock R (300nM) 5μM stock	3.6 3.6	F (300nM) 5μM stock R (600nM) 5μM stock	3.6 7.2	F (300nM) 5μM stock R (900nM) 5μM stock	3.6 10.8
ddH₂O	15	ddH₂O	16.8	ddH₂O	14.4	ddH₂O	10.8	ddH₂O	7.2
F (600nM) 5μM stock R (50nM) 1μM stock	7.2 3	F (600nM) 5μM stock R (100nM) 5μM stock	7.2 1.2	F (600nM) 5μM stock R (300nM) 5μM stock	7.2 3.6	F (600nM) 5μM stock R (600nM) 5μM stock 7.2	7.2 7.2	F (600nM) 5μM stock R (900nM) 5μM stock	7.2 10.8
ddH₂O	11.4	ddH₂O	13.2	ddH₂O	10.8	ddH₂O		ddH₂O	3.6
F (900nM) 5μM stock R (50nM) 1μM stock	10.8 3	F (900nM) 5μM stock R (100nM) 5μM stock	10.8 1.2	F (900nM) 5μM stock R (300nM) 5μM stock	10.8 3.6	F (900nM) 5μM stock R (600nM) 5μM stock	10.8 7.2	F (900nM) 5μM stock R (900nM) 5μM stock	10.8 10.8
ddH₂O	7.8	ddH₂O	9.6	ddH₂O	7.2	ddH₂O	3.6	ddH₂O	0
NTC 1 F(50nM) 1μM stock R (50nM) 1μM stock	3 3							NTC 2 F (900nM) 5μM stock R (900nM) 5μM stock	10.8 10.8
ddH₂O	15.6							ddH₂O	0

143

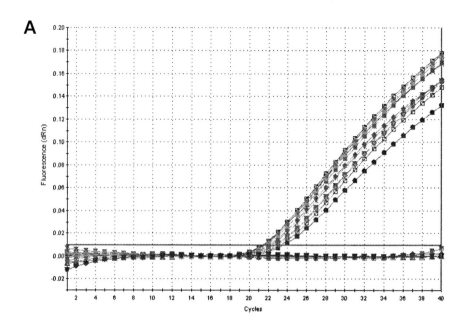

A

B

Forward primer concentration nM	Reverse primer concentration nM			
	100	300	600	900
100	25.01			
100		24.49		
100			23.77	
100				24.23
300	23.13			
300		23.1		
300			22.62	
300				22.58
600	23.14			
600		22.82		
600			23.13	
600				22.68
900	No Ct			
900		No Ct		
900			No Ct	
900				No Ct
ntc	No Ct			

Figure 3. Optimisation of primer pairs. A: Amplification plot of primer optimization assays. The corresponding CTs are presented in B: At 900nM concentrations of forward primer all reactions are inhibited. C: The lowest CT (22.58) and highest ΔRn results from a combination of 300 nM forward primer and 900 nM reverse primer.

144

C

Primer Optimisation Matrix

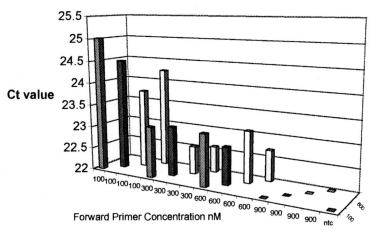

Table 3. Primer optimisation protocol for SYBR Green I reaction. This protocol uses the Stratagene Brilliant SYBR Green I Core reagent kits.

Reagent	Volume	Volume
Master mix	1 x 60μl reaction (μl)	27 x 60μl
[Primers in matrix]	[21.6]	[583.2]
ddH$_2$0	TBD*	TBD*
buffer (10x)	6	162
MgCl$_2$ (50 mM) (2.5 mM final concentration)	3	81
dNTP (20 mM) (0.8 mM final)	2.4	64.8
Rox diluted 1:200	0.9	24.3
Sybr Green I (dilute 1/5000)	0.6	16.2
Taq (5 U/μl)	0.6	16.2
Template (aim for 10^4 to 10^6 copies)	TBD*	TBD*
		Final volume 1620

*TBD, to be determined. The volume of water to be added will depend upon the concentration of the template.
Remember to remove the reaction mix for the NTC before adding template and add water to compensate.

Perform a qPCR assay:
1 cycle: 95°C 10:00
40 cycles: 95°C 00:30
Annealing temp 01:00   Collect data   72°C 00:30   Collect data

Followed by a melting profile:
1 cycle: 95°C 01:00
41 cycles: Collect data
starting at 55°C hold for 00:30 and repeat incrementing temperature by 1°C per cycle.

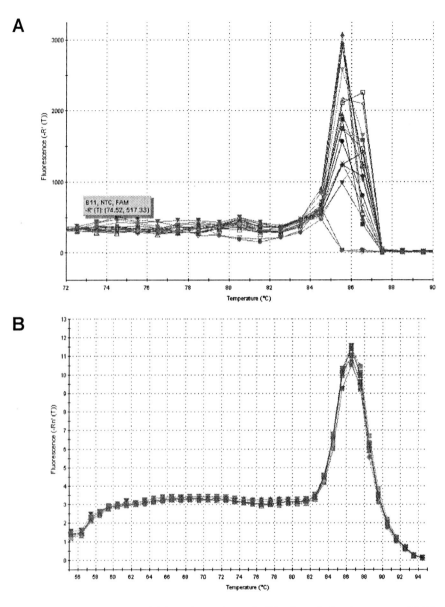

Figure 4. SYBR Green I Melting Curve Data. A. The first derivative view, –R'(T) with respect to temperature provides a clear view of the rate of SYBR Green I loss and the temperature range over which this occurs. For this example a view of that data between 72°C and 90°C will be used. The small peak at 74.5°C is more than likely due to primer dimer product formation since this is the only peak to occur in the NTC sample. The main peaks occur around 85.55°C though there are some with a distinctly different profile and a peak at 86.55°C. These distinct profiles represent different products in the final PCR product. B shows an optimized dissociation curve.

146

In order to perform a melting curve all products from the PCR are held at 95°C to ensure that they are fully separated and then cooled to ensure complete hybridisation. These duplex molecules are then subjected to incubations at increments of temperature. The period of hold and the temperature incremental steps influence the stringency of the data, a longer hold with smaller temperature steps results in a more stringent definition of the melting profile of the product.

The raw data from a melting profile are presented as a first derivative plot of fluorescence units against temperature (Figure 4A) which provides a clear view of the rate of SYBR Green I loss and the temperature range over which this occurs. The small peaks at 74.5°C and 80.5°C are most likely to be due to primer dimer product and non-specific product formation since these are the only peaks to occur in the NTC sample. The main peaks occur at around 85.55°C though there are some reactions with a distinctly different profile and a further peak at 86.55°C. These distinct profiles represent different products in the final PCR product and are clear evidence that this reaction is sub-optimal. Once optimised, the assay looses the additional peaks and now has a single melting profile indicating a single product (Figure 4B).

PCR Amplification Efficiency

Having chosen the optimum concentration for the primers it is important to determine the efficiency of the qPCR reaction. The standard curve is an excellent tool for examining the quality of the overall qPCR assay. This is particularly important if multiple reactions are to be combined in multiplex or if the experiment will be run in the absence of a standard curve and a comparative quantification determination (or $\Delta\Delta C_t$) is to be used. In a perfect world, every amplicon will be replicated perfectly to generate a second copy during every amplification cycle. Assuming that a labelled probe binds to every single amplicon, each amplicon produced will result in one fluorophore molecule being released at each cycle. Hence the critical question concerns the minimum number of molecules of fluorophore that can be detected. If this were one, it would detect one molecule at cycle one. If it were eight molecules that the instrument could detect, it would detect one template input molecule at cycle four and if it could detect only 512 molecules of label, it would take ten cycles to detect a single template input molecule. For most instruments a ballpark figure of 10^{11} free FAM molecules can be detected, with reasonable confidence, above combined background noise. It will require 38 cycles of a perfect serial duplication PCR reaction to produce 1.4×10^{11} copies from an initial single template copy. Therefore, a plot of CT versus the log of the copy number of a utopian qPCR assay has a gradient (slope) of -3.3 and a y-intercept of around 38. For the purposes of this discussion it is worthy of note that the R^2 is indicative of the fit of each data point to the line. A number of factors influence this value; the accuracy of the dilution series, pipetting reproducibility for each data point, contamination or non-specific fluorescence detection, lack of uniformity on the thermal block. With this in mind the gradient, intercept and R^2 values can be used as indicators of the efficiency and sensitivity of the specific qPCR assay (see below).

One Enzyme/One Tube TaqMan RT-PCR Protocol

Preparations

- Maintain a dedicated set of micro pipettes and use filter barrier tips for all qRT-PCR reactions. They must be calibrated regularly, especially pipettes dispensing 10, 2 or 1 µl
- Use RNase free water, aliquot (20ml) and store at –20°C.
- Always aliquot all reaction components and use fresh aliquots if product is detected in no template control (NTC) or contamination is suspected.
- Two NTC reactions, one prepared at the beginning of the assay before any DNA is dispensed, and one at the end should always be performed, to confirm the absence of contamination.
- Defrost all reagents on ice prior to making up reaction mixes. Avoid exposing fluorescent probes and SYBR Green I to light (wrap in tin foil).
- Perform the qRT-PCR as soon as possible.

Primers and Probes

Forward and reverse primers (10 μM each) can be mixed together and should be stored alongside the probes (5 μM) in aliquots at -70 C.

RT-PCR Enzyme

Tth DNA Polymerase 2.5 U/μl

RT-PCR Solutions

- 5 x *Tth* RT-PCR buffer (250 mM Bicine, 575 mM potassium acetate, 0.05 mM EDTA, 300 nM Passive Reference ROX, 40% (w/v) glycerol, pH 8.2),
- 10 mM dATP, 10 mM dCTP, 10 mM dGTP, 20 mM dUTP, 25 mM $Mn(OAc)_2$

The 5x *Tth* buffer contains a passive reference dye that does not participate in the 5'-nuclease assay. It provides an internal reference to which the reporter dye signal is normalised during data analysis. This corrects for fluorescent fluctuations resulting from pipetting errors of the master mix. It does not, of course, correct for pipetting errors of the template. Division of the emission intensity of the reporter dye by the emission intensity of the passive reference generates the normalised reporter (ΔRn). A graph of ΔRn versus cycle number has three stages. Initially ΔRn appears as a flat line because the fluorescent signal is below the detection threshold of the fluorimeter. In the second stage, the signal becomes detectable and increases exponentially. Eventually, a third plateau stage is reached where the ΔRn is linear (Figure 1).

A major advantage of collecting data during every cycle of the PCR reaction is that this provides additional information about the reaction that is most useful when problems arise. All real-time instruments permit a cycle-by-cycle analysis of the raw fluorescence data to show how fluorescence changes over time. The multicomponent view displays the complete spectral contribution of each dye and the

149

background component in a particular well over the duration of the whole run. Problems with the probe, background or even electricity supply are readily detectable.

Preparation of Master Mix

Preparation of a master mix reduces the number of pipetting steps and reagent transfers and so increases the reproducibility of the results. Mix buffer, dNTPs, $Mn(OAc)_2$, primer and probes by briefly vortexing, followed by centrifugation to collect residual liquid from the top and sides of the tubes. Mix enzyme by inverting the tube several times, followed by brief centrifugation. Make the master mix by combining the reagents in the order shown on Table 4, mix gently by repeatedly pipetting up and down (making sure there are no bubbles) and finally add *Tth* polymerase and mix gently again.

Table 4. One enzyme/one tube reagents and amplification set-up.

Reagent	Volume/25 μl reaction (μl)	Final concentration
5X RT-PCR buffer	5.0	1X
25mM $Mn(OAc)_2$	3.0	3 mM
10mM dATP	0.75	300 μM
10mM dCTP	0.75	300 μM
10mM dGTP	0.75	300 μM
20mM dUTP	0.75	600 μM
10 μM Forward Primer	0.5	200 nM
10 μM Reverse Primer	0.5	200 nM
5 μM Probe	0.5	100 nM
Tth DNA Pol (2.5 U/μL)	1.0	0.1 U/μl
RNase free water	to 20	

Perform a qRT-PCR protocol of:

	Step	Time	Temperature	Action
1 cycle:	RT	3 min	60°C	
40 cycles:	Denaturation	15 sec	92°C	
	Annealing/extension	45 sec	62°C	Collect data

Preparation of Standard Curve

The first time an assay is run, a full standard curve experiment must be carried out. Subsequently, only 5 dilution points need to be included with every run.

1. Prepare duplicate 10-fold serial dilutions of the DNA or RNA to be used to generate the standard curve.
2. The dilutions should cover the range 1×10^1-1×10^8 starting copy numbers. CTs for each dilution point will be determined in triplicate; hence each standard curve will be made up of a total of six CTs per dilution point.
3. Aliquot 60 µl of the master mixes into adjacent wells of a 96-well reaction plate. Twelve wells should be reserved for no template controls.
2. Add 15 µl from each of the dilution points to individual sample wells and aspirate using the micropipette tip.
3. To each of the three no template control wells add 15 µl of water.
4. Transfer 25 µl of master mix/template and master mix/water from each well to the empty wells in the two rows below to generate triplicate samples.
5. Cap the wells and ensure that they are properly sealed. It is crucial to ensure that the lids are tightly sealed as the small volumes in which the assays are carried out mean that product is easily lost to evaporation.
6. Transfer the plate to the thermal cycler and perform the RT-PCR reaction as detailed in Table 4.

Template Reaction

1. Aliquot 40 µl of the master mix into the adjacent wells of every other row of a 96-well reaction plate. Four wells should be reserved for no template controls (NTC) and ten wells for the standard curve.
2. Add 10 µl of RNA template to individual unknown sample wells and aspirate using the micropipette tip.
3. To each of the two NTC wells 10 µl of water.
4. Transfer 25 µl of master mix/template and master mix/water from

each well to the empty well in the row below to generate duplicate samples.
5. Cap the wells.
6. Take the ten-fold serial dilutions of the standard 10^2-10^6 copies of amplicon and add 10 µl to each of the five wells containing the master mix.
7. Aspirate using the micropipette tip and transfer 25 µl to each adjacent empty well.
8. Cap the tubes and ensure that they are properly closed.
9. Transfer the plate to the thermal cycler and perform the RT-PCR assay (Table 4).
10. Analyze the results (Figure 5).

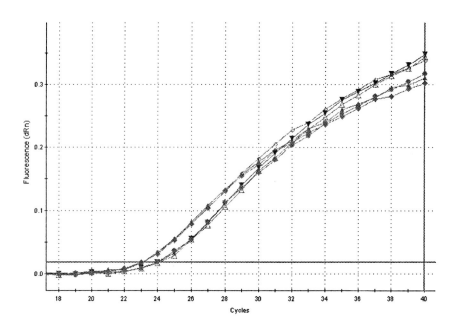

Figure 5. Amplification plots of two templates. Total RNA prepared from colorectal cancer biopsies was analysed by a one enzyme/one tube qRT-PCR assay. Each template was analyzed in triplicate and CTs were converted to copy numbers using an amplicon-specific standard curve coupled to normalization to total RNA quantitated using Ribogreen. The high reproducibility of the assay is apparent, with the 1 CT difference translating into 5,000 ±400 copies of mRNA/µg total RNA and 10,000 ±750 copies of mRNA/µg total RNA.

Two Enzyme/Two Tube TaqMan RT-PCR Protocol

Enzymes

- MMLV-RT 50 U/μl
- AmpliTaq Gold 5 U/μl

The correct choice of DNA polymerase is important as not all possess the 5'–3' nuclease activity necessary for the cleavage of the fluorogenic hybridisation probe bound to its target amplicon (Kreuzer *et al.*, 2000).

Solutions

The two-step RT-PCR reaction requires two reaction mixes:

- RT Reaction Mix
- PCR Reaction Mix

1. RT step
- 2.5 mM dATP, 2.5 mM dCTP, 2.5 mM dGTP, 2.5 mM dTTP,
- 50 μM Oligo d(T)$_{16}$ or 10 μM sequence-specific reverse primer
- 10x RT buffer: 500 mM KCl, 100 mM Tris-HCl, pH 8.3
- 25 mM MgCl$_2$

2. PCR step

- 10x TaqMan Buffer 500 mM KCl, 0.1 mM EDTA, 100 mM Tris-HCl, pH 8.3, and 600 nM Passive Reference
- 10 mM dATP, 10 mM dCTP, 10 mM dGTP, 20 mM dUTP,
- 25 mM MgCl$_2$

Preparation of Master Mixes

Mix the individual ingredients (*except for enzymes*) for the separate RT and PCR steps by briefly vortexing, followed by centrifugation to

Table 5. Two enzyme/two tube reagents and amplification set-up.

RT-Step

Reagent	Volume/12.5 µl reaction (µl)	Final concentration
10 x RT buffer	1.25	1x
25 mM MgCl$_2$	2.75	5.5 mM
2.5 mM dNTP mix	2.50	500 µM each
10 µM specific primer*	0.25	200 nM
50 U/µl MMLV-RT	0.3	1.25 U/µl
RNase-free water	2.95	

Place at 48 °C for 30 min. If using oligo-dT$_{16}$ primer, a 10 min step at 25 °C is required to maximize primer/template annealing.

* If using oligo-dT$_{16}$ primer, the final concentration should be 2.5 µM, *i.e.* use 0.63 µl

PCR-Step

Reagent	Volume/25 µl reaction (µl)	Final concentration
10 x PCR buffer	2.50	1X
25 mM MgCl$_2$	5.5	5.5 mM
10 mM dATP	0.5	200 µM
10 mM dCTP	0.5	200 µM
10 mM dGTP	0.5	200 µM
20 mM dUTP	0.5	400 µM
10 µM Forward Primer	0.5	200 nM
10 µM Reverse Primer	0.5	200 nM
5 µM Probe	0.5	100 nM
AmpliTaq Gold (5 U/µL)	0.125	0.025 U/µl
RNase free water	18.88	
cDNA template	2.5	

Perform a PCR protocol of:

	Step	Time	Temperature	Action
1 cycle:	Activation	10min	95°C	
40 cycles:	Denaturation	15 sec	95°C	
	Annealing/extension	45 sec	62°C	Collect data

collect residual liquid from the top and sides of the tubes. Mix enzyme by inverting the tube several times, followed by brief centrifugation. Make the master mix by combining the reagents in the order shown on Table 5, mix gently by repeatedly pipetting up and down (making sure there are no bubbles) and finally add MMLV-RT and mix gently again. For SYBR Green I assays optimisation for specificity is essential. Therefore, $MgCl_2$ (3 mM instead of 5 mM) and primer (20-50 nM instead of 200 nM) concentrations are lower than in standard assays.

Preparation of Standard Curve

The first time an assay is run, a full standard curve experiment must be carried out. Subsequently, only 5 dilution points need to be included with every run.

1. Prepare duplicate 10-fold serial dilutions of the DNA or RNA to be used to generate the standard curve
2. The dilutions should cover the range $1x10^1$-$1x10^8$ starting copy numbers. CTs for each dilution point will be determined in triplicate, hence each standard curve will be made up of a total of six CTs per dilution point.

RT-Step

3. Prepare the RT reaction mix as described in Table 5. Each assay requires 10 μl of RT-mix and 2.5 μl of RNA. Hence RTs carried out on eight duplicate dilution series will require 10 x 8 x 2 = 160 μl of RT mix. Two NTCs require an additional 20 μl. Therefore, to be on the safe side, 200 μl should be prepared.
4. Aliquot 10 μl of the master mixes into a microfuge tube.
5. Add 2.5 μl from each of the dilution points to individual tubes and aspirate using the micropipette tip.
6. To each of the two NTC tubes add 2.5 μl of water.
7. Cap the tubes and ensure that they are properly sealed.
8. Transfer the plate to a thermal cycler and perform the RT-step reaction as detailed in Table 5.

PCR Step

9. Prepare the PCR Reaction Mix as described in Table 5. Each assay requires 22.5 µl of PCR mix and 2.5 µl of cDNA. Hence PCR assays carried out on eight duplicate dilution series in triplicate will require 22.5 x 8 x 2 x 3 = 1080 µl of PCR mix. Six NTCs require an additional 135 µl. Therefore, to be on the safe side, 1.3 ml should be prepared.
10. Aliquot 60 µl of the master mixes into the adjacent wells of every other row of a 96-well reaction plate.
11. Add 15 µl from each of the dilution points to individual wells and aspirate using the micropipette tip.
12. To each of the two NTC wells add 15 µl of water.
13. Transfer 25µl of master mix/template and master mix/water from each well to the empty wells in the two rows below to generate triplicate samples.
14. Cap the wells and ensure that they are properly sealed. It is crucial to ensure that the lids are tightly sealed as the small volumes in which the assays are carried out mean that product is easily lost to evaporation.
15. Transfer the plate to the thermal cycler and perform the PCR reaction as detailed in Table 5.

Unknowns Reaction

RT Step

1. Prepare the RT reaction mix as described in Table 5. Each assay requires 10 µl of RT-mix and 2.5 µl of RNA. The exact amount of RT master mix required will depend on the number of assays. Two NTCs require an additional 20 µl.
2. Aliquot 10 µl of the master mixes into a microfuge tube.
3. Add 2.5 µl (5-200 ng RNA) to individual tubes and aspirate using the micropipette tip.
4. To each of the two NTC tubes add 2.5 µl of water.
5. Cap the tubes and ensure that they are properly sealed.
6. Transfer the plate to a thermal cycler and perform the RT step reaction as detailed in Table 5.

PCR Step

7. Prepare the PCR master mix as described in Table 5. Each assay requires 22.5 µl of PCR master mix and 2.5 µl of cDNA. The exact amount of PCR master mix required will depend on the number of assays. Allow for four NTC wells
8. Aliquot 45 µl of the PCR master mixes into the adjacent wells of every other row of a 96-well reaction plate.
9. Add 5 µl from each cDNA preparation to individual wells and aspirate using the micropipette tip.
10. To each of two NTC wells add 5 µl of sham cDNA, to a further two NTC wells add 5 µl H_2O.
11. Transfer 25µl of master mix/template and master mix/water from each well to the empty wells in the row below to generate duplicate samples.
12. Cap the tubes and ensure that they are properly sealed.
13. Transfer the plate to a real-time thermal cycler and perform the PCR step as detailed in Table 5.

Trouble Shooting

Two positive controls should always be included, one for the RT step and one for the PCR step. The controls should be RNA or DNA that undergo RT-PCR or PCR, respectively. If no fluorescence is detected with the RT control, but the PCR control is normal, then there was a problem with the RT step. If neither result in fluorescence, then the problem was with the PCR.

RT Step

• Check that all reagents were mixed properly and have been added at the correct concentration.
• If using random primers (*not advisable!*) or oligo-dT, was a 10 min step at 25°C step included?
• If using specific primers, was the reverse primer used?
• Was the temperature of the RT reaction too high? If using 37°C

or 42°C, the temperature can be increased to overcome secondary structure problems. However, this will reduce the activity of the RT.

- If the target is very GC rich or is known to form extensive secondary structures, denature RNA/primer mix *(without enzyme)* for 5 min at 75°C.
- For such templates it may also be useful to increase the RT reaction time
- Reduce/increase the amount of RNA.
- Use 5–20 U of an RNase inhibitor, especially if the assay contains low amounts of RNA (less than 10 ng).

PCR Step

Observation: No increase in fluorescence with cycling.

Possible causes:

- SYBR Green I: Ensure the correct dilution of SYBR Green I was used. Is the SYBR Green I concentration too high? Once SYBR Green I has been diluted it goes off very quickly and should be kept in the dark at 5°C for no more than one week. Dilution into a solution of DMSO may stabilize SYBR Green I, but may also affect the PCR assay. One major consideration when storing working solutions of SYBR stains is that there is a significant increase in the pH of Tris buffers when stored at 4°C versus room temperature. If buffers were prepared at pH 8.0 at room temperature, then the pH will increase to 8.5 at 4°C. This increased pH is beyond the range at which SYBR stains are most stable.
- TaqMan: The probe is not binding to the target efficiently because the annealing temperature is too high. Verify the calculated Tm using appropriate software. Note that Primer Express Tms can be significantly different than Tms calculated using other software packages.
- Molecular Beacon/Scorpion: The probe is not binding to the target efficiently because the loop portion (Molecular Beacon) or the probe (Scorpion) is not completely complementary to the target.

Perform a melting curve analysis to determine if the probe binds to a perfectly complementary target. Make sure the Scorpion probe sequence is complementary to the newly synthesised strand. The assay medium may contain insufficient salt for MB stems to form.
- A reagent is missing from the PCR reaction, repeat the PCR.
- The $MgCl_2$ concentration is not optimal. Increase it in 0.5 mM increments.
- Hot start DNA polymerase was not activated. Ensure that the 10 minute. incubation at 95°C was performed as part of the cycling parameters.
- Ensure the annealing and extension times are sufficient. Check the length of the amplicon and increase the extension time if necessary.
- The probe is not binding to the target efficiently because the PCR product is too long. Design the primers so that the PCR product is <150 bp in length.
- The probe is not binding to the target efficiently because the Mg^{2+} concentration is too low. Perform a Mg^{2+} titration to optimise the concentration.
- The probe has a non-functioning fluorophore. Verify that the fluorophore functions by detecting an increase in fluorescence in the denaturation step of thermal cycling or at high temperatures in a melting curve analysis. If there is no increase in fluorescence, re-synthesise the probe.
- The reaction is not optimised and no or insufficient product is formed. Verify formation of enough specific product by gel electrophoresis.
- Ensure that not more than 10% (v/v) cDNA was used in the PCR step.

Analysis

Quantitative RT-PCR data can be expressed relative to an internal standard ("relative quantification") (Fink *et al.*, 1998) or an external standard curve ("absolute" (Bustin, 2000) or "standard curve quantification" (Ginzinger, 2002)").

Relative Quantification

Relative quantification determines the changes in steady-state mRNA levels of a gene relative to the levels of an internal control RNA (Fink *et al.*, 1998). This reference is usually a housekeeping (HK) gene that is either co-amplified in the same tube or amplified in a separate tube. Therefore, relative quantification does not require standards with known concentrations and all samples are expressed as an n-fold difference relative to the HK gene and the number of target gene copies are normalised to the HK gene. In theory, this should be superior to and far more convenient than absolute quantification. This is because the result is a ratio, hence RNA concentration is irrelevant and there are several mathematical models that calculate the relative expression ratio, some of which correct for differences in amplification efficiency (Liu *et al.*, 2002a; Liu *et al.*, 2002b; Meijerink *et al.*, 2001; Peccoud *et al.*, 1996; Pfaffl, 2001; Pfaffl *et al.*, 2002; Soong *et al.*, 2001) while some do not (Livak *et al.*, 2001; Winer *et al.*, 1999).

The two main problems with relative quantification are that (1) this approach tends to introduce a significant statistical bias that results in misleading biological interpretation (Hocquette *et al.*, 2002). This is particularly true when there are vast differences in the expression levels of target and normaliser or when the target gene is expressed at very low levels when the relationship between the two may not be linear; (2) it is difficult to find a HK gene whose expression is constant and against which the target gene copy numbers can be normalized during the experimental conditions. This is particularly so for *in vivo* biopsies.

GAPDH, β-actin (Kreuzer *et al.*, 1999), histone H3 (Kelley *et al.*, 1993), ribosomal highly basic 23-kDa protein (Jesnowski *et al.*, 2002), cyclophilin (Haendler *et al.*, 1987), β-2-microglobulin (Lupberger *et al.*, 2002) and porphobilinogen deaminase (Fink *et al.*, 1998) are commonly used normalisers for relative quantification. GAPDH was a curious choice, since its gene product is well known to play a role in membrane fusion, microtubule bundling, phosphotransferase activity, nuclear RNA export, DNA replication, DNA repair, apoptosis, age-related neurodegenerative disease, prostate cancer and viral pathogenesis (Sirover, 1999). β-actin should also not be used to normalize qRT-PCR data (Weisinger *et al.*, 1999) as its mRNA levels can vary widely in response to experimental manipulation in human breast epithelial cells (Spanakis, 1993), and blastomeres (Krussel *et al.*, 1998), as well as in various porcine tissues (Foss *et al.*, 1998) and canine myocardium (Carlyle *et al.*, 1996). Matrigel, which is widely used for cell attachment and to induce cell differentiation, adversely affects β-actin mRNA levels (Selvey *et al.*, 2001). A recent systematic analysis and comparison of their usefulness on *in vivo* tissue biopsies has concluded that a single housekeeping gene should not be used for normalisation (Tricarico *et al.*, 2002). The recent demonstration of the effectiveness of normalisation against the geometric mean of multiple carefully selected HK genes is interesting (Vandesompele *et al.*, 2002b). However, this method requires extensive practical validation to identify a combination of reference genes appropriate for every individual experiment, something that is not at all trivial. In addition, as the choice of HK gene panel is tissue or cell-dependent, this is not a universal method. If HK genes are to be used, they must be validated for the specific experimental setup and it is probably necessary to choose more than one, as was done for example for expression profiling of T-helper cell differentiation (Hamalainen *et al.*, 2001).

Absolute Quantification

Absolute quantification provides a more accurate and reliable, albeit more labour-intensive method for the quantification of nucleic acids (Ke *et al.*, 2000b). CTs obtained from an unknown sample are compared to CTs generated from a series of samples of known concentration or copy number. Results can be expressed as copy number per unit mass, *e.g.* μg total RNA. The expression of a target nucleic acid is usually compared across many samples, often from different individuals and sometimes from different tissues. Since small differences in nucleic acid input can lead to large differences in PCR product yield, the amount of starting material must be quantitated with rigorous accuracy to normalize sample data and correct for tube-to-tube differences. Therefore, if absolute quantification is to be accurate, it needs to take that variability into account.

The initial dilutions to use for the standard curve should encompass as large a range as possible and so for optimisation purposes a three-fold to ten-fold dilution series over five to seven orders of magnitude usually serves best. Each sample should be run in triplicate. An example of an ideal standard curve is shown on Figure 6A. A three-fold serial dilution was used and the gradient of −3.358 and 98.5% efficiency indicates that this is a well optimised assay and suitable for application to experimental samples and quantification. In contrast a less than ideal example is shown on Figure 6B. This standard curve was constructed from a two-fold serial dilution of template. The gradient of −4.458 and efficiency 67.6% indicates that the reaction is very inefficient. It would not be advisable to use this assay for quantitative studies and would require a process of trouble shooting and possible redesign before proceeding.

The steps involved in constructing a standard curve are very straightforward:

1. Five to six serial 10-fold dilutions of known concentration or copy number are prepared from whatever standard is being used, *e.g.* amplicon-specific sense-strand oligonucleotides, T7-transcribed RNA, linear plasmid DNA etc.
2. Serial dilutions are analyzed by qRT-PCR in separate sample wells but within the same run and the resulting CTs are recorded.
3. A plot of CT versus logarithm of concentration or the copy number results in a straight line, the standard curve, which is the linear regression line through the data points.
4. The number of target gene copies can then be extrapolated from the standard curve equation: Copy number = (CT) − (intercept) / slope. The replicate readings should be sufficiently close (<0.5 CTs) to indicate a valid analysis. The limits depend on the concentration of the template in the sample. If the readings are too far apart then the sample should be retested.

A standard curve provides several vital pieces of information:

• A high R^2 value (>0.98) indicates that there is near perfect correlation between CT and copy number. Such R^2 values are common for most good, carefully constructed standard curves.

- The slope of the standard curve can be used to determine the efficiency of the PCR reaction by the following equation: Efficiency = $[10^{(-1/\text{slope})}]$-1. In the above example, the efficiency of amplification is 99.5%.

- The y-intercept is less reproducible than the slope, but gives some indication of how sensitive the assay is.

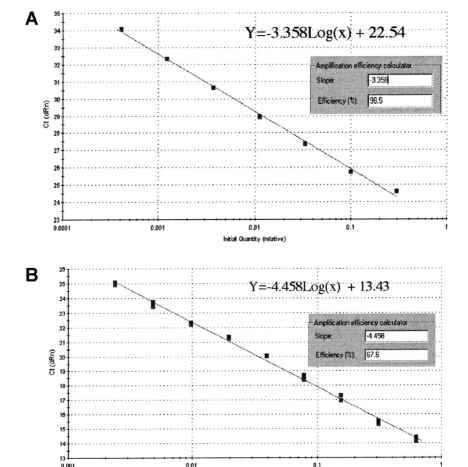

Figure 6. A: Optimised standard curve. B: Poor standard curve.

The main disadvantage of using an external standard is that it cannot provide a control for detecting inhibitors of the PCR reaction. This requires the addition of an internal control template that could be amplified and any variation from the expected CT would suggest the presence of inhibitors. Inclusion of such an internal control generates more confidence in negative results where no template is detectable.

Normalisation

Data normalisation, while a vital aspect of experimental design (Karge *et al.*, 1998), remains a real problem for absolute quantification (Thellin *et al.*, 1999), especially when the samples are from *in vivo* biopsies obtained from different individuals. Ideally, an internal standard should be used which is expressed at a constant level among different tissues, at all stages of development, and which should be unaffected by any experimental treatment. In addition, an endogenous control should also be expressed at roughly the same level as the RNA under study. In the absence of any one single RNA with a constant expression level in all of these situations (Haberhausen *et al.*, 1998), various HK gene, rRNA and total RNA are most commonly used to normalize gene expression patterns.

Tissue Culture

Experiments involving tissue culture cells that are being subjected to certain treatments, with mRNA levels measured before and after treatment can be normalised against cell numbers, with mRNA levels expressed as copy numbers per cell. Assuming that cells are counted accurately, this will generate precise, accurate and meaningful results. Alternatively, it may be not necessary to carry out absolute quantification experiments as is likely that one of the many HK genes proposed as internal standards will be suitable as a normaliser for relative quantification. Indeed, this is one of the few experimental designs where relative quantification may be acceptable. However, note that cellular subpopulations of the same pathological origin can be highly heterogeneous (Goidin *et al.*, 2001) and that careful consideration of the appropriate HK genes is crucial. The mRNA levels of the target are reported as a value relative to the average mRNA levels of the selection of HK genes.

Nucleated Blood Cells (NBC)

Counting NBCs and expressing mRNA copy numbers per cell is the simplest way of reporting mRNA levels. However, this may not be accurate, since blood is made up of numerous subpopulations of cells of different lineage at different stages of differentiation, and differences in mRNA expression patterns are likely to be masked by this variability, a problem exacerbated when attempting to compare mRNA levels between different individuals. Quantification relative to a lineage-specific marker may be appropriate: Primers and fluorescent probes have been reported for numerous subtypes of NBC, *e.g.* CD45 (pan-NBC), CD3 (T-lymphocyte), CD19 (B-lymphocyte), CD14 (monocyte), and CD66 (granulocyte), and the specificity of quantification by real-time RT-PCR compares well with flow cytometric analysis of enriched cell populations (Pennington *et al.*, 2001). Normalisation against total RNA is also possible, as there is relatively little variation in the amount of total RNA per NBC (Bustin, 2000). However, normalisation against total RNA does not overcome the problem of variable subpopulations leading to inappropriate quantification and conclusions.

Solid Tissue Biopsies

Biopsies contain numerous cell types in variable proportions and there is no easy way of sorting or counting them without affecting the expression profile of the sample. Cancers in particular contain not just normal and cancer cells, but there may be several subclones of cancer cells together with stromal, immune and vascular cells (Baisse *et al.*, 2001). This variability means that while it is acceptable to generate qualitative results, there must be a question mark over quantitative data. The use of relative quantification is suspect because the mRNA levels of potential normalisers are not known and their levels may be altered locally by tissue-specific factors. Normalisation against cell number is a useful option only for cells obtained from Laser Capture Microdissection (LCM), as it establishes a direct quantitative relationship between the target mRNA copy number and the cells from which the RNA was derived. Normalisation against total cellular RNA (Bustin, 2000) has been shown to produce quantitative results that are biologically meaningful (Tricarico et al., 2002), but is crucially dependent on accurate quantification of the RNA which may not be very reliable. Messenger RNA levels can be normalised against total genomic DNA, with copy numbers expressed as absolute numbers per unit of DNA. Ribosomal RNA (rRNA) has been proposed as an alternative normaliser (Bhatia et al., 1994;Zhong et al., 1999) as rRNA levels vary less under conditions that affect the expression of mRNAs (Barbu et al., 1989; Goidin et al., 2001; Schmittgen et al., 2000). However, rRNA levels are also affected by biological factors and drugs (Spanakis, 1993) and they vary significantly in haemopoietic subpopulations (Raaijmakers et al., 2002) and in both normal and cancer biopsies taken from different individuals (Tricarico et al., 2002). Furthermore, the vast difference in expression levels between rRNA and any target gene can result in misleading quantification data. Normalisation against HK genes runs into the same problems as discussed above with respect to relative quantification and cannot be recommended for RNA extracted from in vivo biopsies.

Quantification of MMP2 and MMP9 mRNA in Colorectal Cancers

The mRNA levels of two metalloproteinases, MMP2 and MMP were quantitated in 39 paired normal and adjacent colorectal cancer tissue biopsies. The probe and primer sequences are shown in Table 6. Two amplicon-specific standard curves were generated as described above (Figure 7) and used to quantitate MMP2 and MMP9 copy numbers in the colorectal biopsies. mRNA copy numbers were normalised against total RNA that was quality assessed using the Agilent 2100 Bioanalyzer and quantitated using Ribogreen assays. The results (Figure 8) suggest that (1) there is a wide range of mRNA levels (greater than 4 orders of magnitude) between different individuals; (2) MMP2 mRNA levels

Table 6. MMP primer and probe sequences.

MMP2	F:	5′-CGCTCAGATCCGTGGTGAG-3′
	R:	5′-TGTCACGTGGCGTCACAGT-3′
	Probe:	5′-TTCTTCTTCAAGGACCGGTTCATTTGGC-3′
MMP9	F:	5′-CCCTGGAGACCTGAGAACCA-3′
	R:	5′-CCCGAGTGTAACCATAGCGG-3′
	Probe:	5′-CCGACAGGCAGCTGGCAGAGGAAT-3′

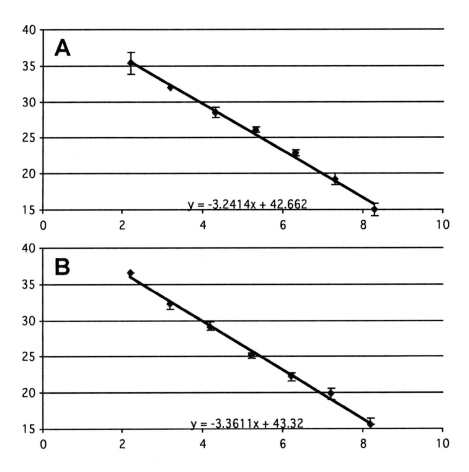

Figure 7. Standard curves for A: MMP2 and B: MMP9. Note that the y-intercept of both standard curves is >40, suggesting that the primer/probe combination is not optimised for sensitivity. However, the slopes indicate that the amplification efficiencies are close to 100%.

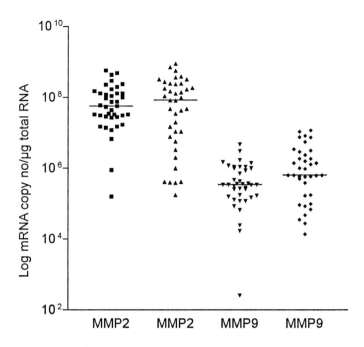

Figure 8. Comparison of MMP2 and MMP9 mRNA copy numbers in adjacent normal and paired colorectal cancer samples.

are higher than those of MMP9 and (3) there is no difference in their expression levels between adjacent normal and cancer tissue.

Conclusions

There can be little doubt that the rapid and easy acquisition of quantitative mRNA expression data by real-time fluorescence-based RT-PCR has placed this technology at the centre of a revolution that is allowing researchers to probe cellular expression profiles with a very high degree of precision. However, the huge variability of samples and templates, the wide range of enzymes, primer, amplicon and chemistry combinations makes a comparison between results obtained in different laboratories difficult. The aim of this chapter was to delineate a series of steps that must be taken if the vast amount of quantitative data generated by qRT-PCR assays is to be not just precise but also biologically relevant.

Acknowledgement

SAB would like to thank Bowel and Cancer Research for supporting some of the research discussed in this chapter.

References

Aatsinki, J.T., Lakkakorpi, J.T., Pietila, E.M., and Rajaniemi, H.J. 1994. A coupled one-step reverse transcription PCR procedure for generation of full-length open reading frames. Biotechniques 16: 282-288.

Akane, A., Matsubara, K., Nakamura, H., Takahashi, S., and Kimura, K. 1994. Identification of the heme compound copurified with deoxyribonucleic acid (DNA) from bloodstains, a major inhibitor of polymerase chain reaction (PCR) amplification. J. Forensic Sci. 39: 362-372.

Al Soud, W.A. and Radstrom, P. 1998. Capacity of nine thermostable DNA polymerases to mediate DNA amplification in the presence of PCR-inhibiting samples. Appl. Environ. Microbiol. 64: 3748-3753.

Al Soud, W.A. and Radstrom, P. 2001. Purification and characterization of PCR-inhibitory components in blood cells. J. Clin. Microbiol. 39: 485-493.

Apfalter, P., Blasi, F., Boman, J., Gaydos, C.A., Kundi, M., Maass, M., Makristathis, A., Meijer, A., Nadrchal, R., Persson, K., Rotter, M.L., Tong, C.Y., Stanek, G., and Hirschl, A.M. 2001. Multicenter comparison trial of DNA extraction methods and PCR assays for detection of *Chlamydia pneumoniae* in endarterectomy specimens. J. Clin. Microbiol. 39: 519-524.

Baisse, B., Bouzourene, H., Saraga, E.P., Bosman, F.T., and Benhattar, J. 2001. Intratumor genetic heterogeneity in advanced human colorectal adenocarcinoma. Int. J. Cancer 93: 346-352.

Barbu, V. and Dautry, F. 1989. Northern blot normalization with a 28S rRNA oligonucleotide probe. Nucleic Acids Res. 17: 7115.

Barnes, W.M. 1994. PCR amplification of up to 35-kb DNA with high fidelity and high yield from lambda bacteriophage templates. Proc. Natl. Acad. Sci. U.S.A 91: 2216-2220.

Baskaran, N., Kandpal, R.P., Bhargava, A.K., Glynn, M.W., Bale, A., and Weissman, S.M. 1996. Uniform amplification of a mixture of deoxyribonucleic acids with varying GC content. Genome Res. 6: 633-638.

Bebenek, K., Abbotts, J., Roberts, J.D., Wilson, S.H., and Kunkel, T.A. 1989. Specificity and mechanism of error-prone replication by human immunodeficiency virus-1 reverse transcriptase. J. Biol.Chem. 264: 16948-16956.

Belec, L., Authier, J., Eliezer-Vanerot, M.C., Piedouillet, C., Mohamed, A.S., and Gherardi, R.K. 1998. Myoglobin as a polymerase chain reaction (PCR) inhibitor: a limitation for PCR from skeletal muscle tissue avoided by the use of *Thermus thermophilus* polymerase. Muscle Nerve 21: 1064-1067.

Beutler, E., Gelbart, T., and Kuhl, W. 1990. Interference of heparin with the polymerase chain reaction. Biotechniques 9: 166.

Bhatia, P., Taylor, W.R., Greenberg, A.H., and Wright, J.A. 1994. Comparison of glyceraldehyde-3-phosphate dehydrogenase and 28S- ribosomal RNA gene expression as RNA loading controls for northern blot analysis of cell lines of varying malignant potential. Anal. Biochem. 216: 223-226.

Bock, O., Kreipe, H., and Lehmann, U. 2001. One-step extraction of RNA from archival biopsies. Anal. Biochem. 295: 116-117.

Bomjen, G., Raina, A., Sulaiman, I.M., Hasnain, S.E., and Dogra, T.D. 1996. Effect of storage of blood samples on DNA yield, quality and fingerprinting: a forensic approach. Indian J. Exp. Biol. 34: 384-386.

Brooks, E.M., Sheflin, L.G., and Spaulding, S.W. 1995. Secondary structure in the 3' UTR of EGF and the choice of reverse transcriptases affect the detection of message diversity by RT-PCR. Biotechniques 19: 806-815.

Buell, G.N., Wickens, M.P., Payvar, F., and Schimke, R.T. 1978. Synthesis of full length cDNAs from four partially purified oviduct mRNAs. J. Biol. Chem. 253: 2471-2482.

Burchill, S.A., Lewis, I.J., and Selby, P. 1999. Improved methods using the reverse transcriptase polymerase chain reaction to detect tumour cells. Br. J. Cancer 79: 971-977.

Burgos, J., Ramirez, C., Tenorio, R., Sastre, I., and Bullido, M. 2002. Influence of reagents formulation on real-time PCR parameters. Mol. Cell. Probes 16: 257.

Bustin, S.A. 2000. Absolute quantification of mRNA using real-time reverse transcription polymerase chain reaction assays. J. Mol. Endocrinol. 25: 169-193.

Bustin, S.A. 2002. Quantification of mRNA using real-time reverse transcription PCR (RT-PCR): trends and problems. J. Mol. Endocrinol 29: 23-39.

Bustin, S.A. and Dorudi, S. 1998. Molecular assessment of tumour stage and disease recurrence using PCR- based assays. Mol. Med. Today 4: 389-396.

Carlyle, W.C., Toher, C.A., Vandervelde, J.R., McDonald, K.M., Homans, D.C., and Cohn, J.N. 1996. Changes in beta-actin mRNA expression in remodeling canine myocardium. J. Mol. Cell. Cardiol. 28: 53-63.

Carninci, P., Nishiyama, Y., Westover, A., Itoh, M., Nagaoka, S., Sasaki, N., Okazaki, Y., Muramatsu, M., and Hayashizaki, Y. 1998. Thermostabilization and thermoactivation of thermolabile enzymes by trehalose and its application for the synthesis of full length cDNA. Proc. Natl. Acad. Sci. U.S.A. 95: 520-524.

Chakrabarti, R. and Schutt, C.E. 2001. The enhancement of PCR amplification by low molecular weight amides. Nucleic Acids Res. 29: 2377-2381.

Chakrabarti, R. and Schutt, C.E. 2002. Novel sulfoxides facilitate GC-rich template amplification. Biotechniques 32: 866, 868, 870-2, 874.

Chen, Z., Swisshelm, K., and Sager, R. 1994. A cautionary note on reaction tubes for differential display and cDNA amplification in thermal cycling. Biotechniques 16: 1002-4, 1006.

Chevet, E., Lemaitre, G., and Katinka, M.D. 1995. Low concentrations of tetramethylammonium chloride increase yield and specificity of PCR. Nucleic Acids Res. 23: 3343-3344.

Chiocchia, G. and Smith, K.A. 1997. Highly sensitive method to detect mRNAs in individual cells by direct RT-PCR using Tth DNA polymerase. Biotechniques 22: 312-318.

Cohen, C.D., Grone, H.J., Grone, E.F., Nelson, P.J., Schlondorff, D., and Kretzler, M. 2002. Laser microdissection and gene expression analysis on formaldehyde-fixed archival tissue. Kidney Int. 61: 125-132.

Comey, C.T., Jung, J.M., and Budowle, B. 1991. Use of formamide to improve amplification of HLA DQ alpha sequences. Biotechniques 10: 60-61.

Cone, R.W., Hobson, A.C., and Huang, M.L. 1992. Coamplified positive control detects inhibition of polymerase chain reactions. J. Clin. Microbiol. 30: 3185-3189.

Coombs, N.J., Gough, A.C., and Primrose, J.N. 1999. Optimisation of DNA and RNA extraction from archival formalin-fixed tissue. Nucleic Acids Res. 27: e12.

Curry, J., McHale, C., and Smith, M.T. 2002. Low efficiency of the Moloney murine leukemia virus reverse transcriptase during reverse transcription of rare t(8;21) fusion gene transcripts. Biotechniques 32: 768, 770, 772, 754-755.

Cusi, M.G., Valassina, M., and Valensin, P.E. 1994. Comparison of M-MLV reverse transcriptase and Tth polymerase activity in RT-PCR of samples with low virus burden. Biotechniques 17: 1034-1036.

de Vries, T.J., Fourkour, A., Punt, C.J., Ruiter, D.J., and van Muijen, G.N. 2000. Analysis of melanoma cells in peripheral blood by reverse transcription-polymerase chain reaction for tyrosinase and MART-1 after mononuclear cell collection with cell preparation tubes: a comparison with the whole blood guanidinium isothiocyanate RNA isolation method. Melanoma Res. 10: 119-126.

Demeke, T. and Adams, R.P. 1992. The effects of plant polysaccharides and buffer additives on PCR. Biotechniques 12: 332-334.

Easton, L.A., Vilcek, S., and Nettleton, P.F. 1994. Evaluation of a 'one tube' reverse transcription-polymerase chain reaction for the detection of ruminant pestiviruses. J Virol. Methods 50: 343-348.

Edwards, M.C. and Gibbs, R.A. 1994. Multiplex PCR: advantages, development, and applications. PCR Methods Appl. 3: S65-S75.

Fehlmann, C., Krapf, R., and Solioz, M. 1993. Reverse transcriptase can block polymerase chain reaction. Clin. Chem. 39: 368-369.

Fink, L., Kinfe, T., Stein, M.M., Ermert, L., Hanze, J., Kummer, W., Seeger, W., and Bohle, R.M. 2000. Immunostaining and laser-assisted cell picking for mRNA analysis. Lab. Invest. 80: 327-333.

Fink, L., Seeger, W., Ermert, L., Hanze, J., Stahl, U., Grimminger, F., Kummer, W., and Bohle, R.M. 1998. Real-time quantitative RT-PCR after laser-assisted cell picking. Nat. Med. 4: 1329-1333.

Foss, D.L., Baarsch, M.J., and Murtaugh, M.P. 1998. Regulation of hypoxanthine phosphoribosyltransferase, glyceraldehyde-3-phosphate dehydrogenase and beta-actin mRNA expression in porcine immune cells and tissues. Anim. Biotechnol. 9: 67-78.

Fredricks, D.N. and Relman, D.A. 1998. Improved amplification of microbial DNA from blood cultures by removal of the PCR inhibitor sodium polyanetholesulfonate. J. Clin. Microbiol. 36: 2810-2816.

Freeman, W.M., Vrana, S.L., and Vrana, K.E. 1996. Use of elevated reverse transcription reaction temperatures in RT-PCR. Biotechniques 20: 782-783.

Freeman, W.M., Walker, S.J., and Vrana, K.E. 1999. Quantitative RT-PCR: pitfalls and potential. Biotechniques 26: 112-115.

Gerard, C.J., Olsson, K., Ramanathan, R., Reading, C., and Hanania, E.G. 1998. Improved quantitation of minimal residual disease in multiple myeloma using real-time polymerase chain reaction and plasmid-DNA complementarity determining region III standards. Cancer Res. 58: 3957-3964.

Gerard, G.F., Fox, D.K., Nathan, M., and D'Alessio, J.M. 1997. Reverse transcriptase. The use of cloned Moloney murine leukemia virus reverse transcriptase to synthesise DNA from RNA. Mol. Biotechnol. 8: 61-77.

Ginzinger, D.G. 2002. Gene quantification using real-time quantitative PCR: an emerging technology hits the mainstream. Exp. Hematol. 30: 503-512.

Goblet, C., Prost, E., and Whalen, R.G. 1989. One-step amplification of transcripts in total RNA using the polymerase chain reaction. Nucleic Acids Res. 17: 2144.

Godfrey, T.E., Kim, S.H., Chavira, M., Ruff, D.W., Warren, R.S., Gray, J.W., and Jensen, R.H. 2000. Quantitative mRNA expression analysis from formalin-fixed, paraffin- embedded tissues using 5' nuclease quantitative reverse transcription- polymerase chain reaction. J. Mol. Diagn. 2: 84-91.

Goidin, D., Mamessier, A., Staquet, M.J., Schmitt, D., and Berthier-Vergnes, O. 2001. Ribosomal 18S RNA Prevails over Glyceraldehyde-3-Phosphate Dehydrogenase and beta-Actin Genes as Internal Standard for Quantitative Comparison of mRNA Levels in Invasive and Non invasive Human Melanoma Cell Subpopulations. Anal. Biochem. 295: 17-21.

Golenberg, E.M., Bickel, A., and Weihs, P. 1996. Effect of highly fragmented DNA on PCR. Nucleic Acids Res. 24: 5026-5033.

Haberhausen, G., Pinsl, J., Kuhn, C.C., and Markert-Hahn, C. 1998. Comparative study of different standardization concepts in quantitative competitive reverse transcription-PCR assays. J. Clin. Microbiol. 36: 628-633.

Haendler, B., Hofer-Warbinek, R., and Hofer, E. 1987. Complementary DNA for human T-cell cyclophilin. EMBO J. 6: 947-950.

Halford, W.P., Falco, V.C., Gebhardt, B.M., and Carr, D.J. 1999. The inherent quantitative capacity of the reverse transcription-polymerase chain reaction. Anal. Biochem. 266: 181-191.

Hall, L.M., Slee, E., and Jones, D.S. 1995. Overcoming polymerase chain reaction inhibition in old animal tissue samples using ethidium bromide. Anal. Biochem. 225: 169-172.

Hamalainen, H.K., Tubman, J.C., Vikman, S., Kyrola, T., Ylikoski, E., Warrington, J.A., and Lahesmaa, R. 2001. Identification and validation of endogenous reference genes for expression profiling of T helper cell differentiation by quantitative real-time RT-PCR.

Anal. Biochem. 299: 63-70.

Harrell, R.A. and Hart, R.P. 1994. Rapid preparation of *Thermus flavus* DNA polymerase. PCR Methods Appl. 3: 372-375.

Henke, W., Herdel, K., Jung, K., Schnorr, D., and Loening, S.A. 1997. Betaine improves the PCR amplification of GC-rich DNA sequences. Nucleic Acids Res. 25: 3957-3958.

Henley, W.N., Schuebel, K.E., and Nielsen, D.A. 1996. Limitations imposed by heteroduplex formation on quantitative RT-PCR. Biochem. Biophys. Res. Commun. 226: 113-117.

Hill, C.S. and Treisman, R. 1995. Transcriptional regulation by extracellular signals: mechanisms and specificity. Cell. 80: 199-211.

Hocquette, J.F. and Brandstetter, A.M. 2002. Common practice in molecular biology may introduce statistical bias and misleading biological interpretation. J. Nutr. Biochem. 13: 370-377.

Hottiger, T., De Virgilio, C., Hall, M.N., Boller, T., and Wiemken, A. 1994. The role of trehalose synthesis for the acquisition of thermotolerance in yeast. II. Physiological concentrations of trehalose increase the thermal stability of proteins *in vitro*. Eur. J Biochem. 219: 187-193.

Jesnowski, R., Backhaus, C., Ringel, J., and Lohr, M. 2002. Ribosomal Highly Basic 23-kDa Protein as a Reliable Standard for Gene Expression Analysis. Pancreatology. 2: 421-424.

Johnson, P.W., Swinbank, K., MacLennan, S., Colomer, D., Debuire, B., Diss, T., Gabert, J., Gupta, R.K., Haynes, A., Kneba, M., Lee, M.S., Macintyre, E., Mensink, E., Moos, M., Morgan, G.J., Neri, A., Johnson, A., Reato, G., Salles, G., van't Veer, M.B., Zehnder, J.L., Zucca, E., Selby, P.J., and Cotter, F.E. 1999. Variability of polymerase chain reaction detection of the bcl-2-IgH translocation in an international multicentre study. Ann. Oncol. 10: 1349-1354.

Jung, R., Soondrum, K., and Neumaier, M. 2000. Quantitative PCR. Clin. Chem.Lab Med. 38: 833-836.

Kainz, P. 2000. The PCR plateau phase - towards an understanding of its limitations. Biochim. Biophys. Acta 1494: 23-27.

Kappes, J.C., Saag, M.S., Shaw, G.M., Hahn, B.H., Chopra, P., Chen, S., Emini, E.A., McFarland, R., Yang, L.C., Piatak, M., Jr., and . 1995. Assessment of antiretroviral therapy by plasma viral load testing: standard and ICD HIV-1 p24 antigen and viral RNA (QC-PCR) assays compared. J. Acquir. Immune. Defic. Syndr. Hum. Retrovirol. 10: 139-149.

Karge, W.H., Schaefer, E.J., and Ordovas, J.M. 1998. Quantification of mRNA by polymerase chain reaction (PCR) using an internal standard and a non-radioactive detection method. Methods Mol. Biol. 110: 43-61.

Ke, D., Menard, C., Picard, F.J., Boissinot, M., Ouellette, M., Roy, P.H., and Bergeron, M.G. 2000a. Development of Conventional and Real-Time PCR Assays for the Rapid Detection of Group B Streptococci. Clin. Chem. 46: 324-331.

Ke, L.D., Chen, Z., and Yung, W.K. 2000b. A reliability test of standard-based quantitative PCR: exogenous vs endogenous standards. Mol. Cell. Probes 14: 127-135.

Keilholz, U., Willhauck, M., Rimoldi, D., Brasseur, F., Dummer, W., Rass, K., de Vries, T., Blaheta, J., Voit, C., Lethe, B., and Burchill, S. 1998. Reliability of reverse transcription-polymerase chain reaction (RT-PCR)- based assays for the detection of circulating tumour cells: a quality- assurance initiative of the EORTC Melanoma Cooperative Group. Eur. J Cancer 34: 750-753.

Kelley, M.R., Jurgens, J.K., Tentler, J., Emanuele, N.V., Blutt, S.E., and Emanuele, M.A. 1993. Coupled reverse transcription-polymerase chain reaction (RT-PCR) technique is comparative, quantitative, and rapid: uses in alcohol research involving low abundance mRNA species such as hypothalamic LHRH and GRF. Alcohol 10: 185-189.

Kotewicz, M.L., Sampson, C.M., D'Alessio, J.M., and Gerard, G.F. 1988. Isolation of cloned Moloney murine leukemia virus reverse transcriptase lacking ribonuclease H activity. Nucleic Acids Res. 16: 265-277.

Kreader, C.A. 1996. Relief of amplification inhibition in PCR with bovine serum albumin or T4 gene 32 protein. Appl. Environ.

Microbiol. 62: 1102-1106.

Kreuzer, K.A., Bohn, A., Lass, U., Peters, U.R., and Schmidt, C.A. 2000. Influence of DNA polymerases on quantitative PCR results using TaqMan™ probe format in the LightCycler (TM) instrument. Mol.Cell. Probes 14: 57-60.

Kreuzer, K.A., Lass, U., Landt, O., Nitsche, A., Laser, J., Ellerbrok, H., Pauli, G., Huhn, D., and Schmidt, C.A. 1999. Highly sensitive and specific fluorescence reverse transcription-PCR assay for the pseudogene-free detection of beta-actin transcripts as quantitative reference. Clin. Chem. 45: 297-300.

Krussel, J.S., Huang, H.Y., Simon, C., Behr, B., Pape, A.R., Wen, Y., Bielfeld, P., and Polan, M.L. 1998. Single blastomeres within human preimplantation embryos express different amounts of messenger ribonucleic acid for beta-actin and interleukin-1 receptor type I. J. Clin. Endocrinol. Metab 83: 953-959.

Kwok, S. and Higuchi, R. 1989. Avoiding false positives with PCR. Nature 339: 237-238.

Lau, J.Y., Qian, K.P., Wu, P.C., and Davis, G.L. 1993. Ribonucleotide vanadyl complexes inhibit polymerase chain reaction. Nucleic Acids Res. 21: 2777.

Lee, K.H., McKenna, M.J., Sewell, W.F., and Ung, F. 1997. Ribonucleases may limit recovery of ribonucleic acids from archival human temporal bones. Laryngoscope. 107: 1228-1234.

Lehmann, U. and Kreipe, H. 2001. Real-time PCR analysis of DNA and RNA extracted from formalin-fixed and paraffin-embedded biopsies. Methods. 25: 409-418.

Lekanne Deprez, R.H., Fijnvandraat, A.C., Ruijter, J.M., and Moorman, A.F. 2002. Sensitivity and accuracy of quantitative real-time polymerase chain reaction using SYBR green I depends on cDNA synthesis conditions. Anal. Biochem. 307: 63-69.

Liu, W. and Saint, D.A. 2002a. A new quantitative method of real time reverse transcription polymerase chain reaction assay based on simulation of polymerase chain reaction kinetics. Anal. Biochem. 302: 52-59.

Liu, W. and Saint, D.A. 2002b. Validation of a quantitative method for real time PCR kinetics. Biochem. Biophys. Res. Commun. 294: 347-353.

Livak, K.J. and Schmittgen, T.D. 2001. Analysis of relative gene expression data using real-time quantitative PCR and the 2(-Delta Delta C(T)) Method. Methods. 25: 402-408.

Louwrier, A. and van der VAlk, A. 2001. Can sucrose affect polymerase chain reaction product formation? Biotechnol. Lett. 23: 175-178.

Lupberger, J., Kreuzer, K.A., Baskaynak, G., Peters, U.R., le Coutre, P., and Schmidt, C.A. 2002. Quantitative analysis of beta-actin, beta-2-microglobulin and porphobilinogen deaminase mRNA and their comparison as control transcripts for RT-PCR. Mol. Cell Probes. 16: 25-30.

Madejon, A., Manzano, M.L., Arocena, C., Castillo, I., and Carreno, V. 2000. Effects of delayed freezing of liver biopsies on the detection of hepatitis C virus RNA strands. J Hepatol. 32: 1019-1025.

Mader, R.M., Schmidt, W.M., Sedivy, R., Rizovski, B., Braun, J., Kalipciyan, M., Exner, M., Steger, G.G., and Mueller, M.W. 2001. Reverse transcriptase template switching during reverse transcriptase-polymerase chain reaction: artificial generation of deletions in ribonucleotide reductase mRNA. J. Lab. Clin. Med. 137: 422-428.

Mahony, J.B., Chong, S., Coombes, B.K., Smieja, M., and Petrich, A. 2000. Analytical sensitivity, reproducibility of results, and clinical performance of five PCR assays for detecting *Chlamydia pneumoniae* DNA in peripheral blood mononuclear cells. J. Clin. Microbiol. 38: 2622-2627.

Malboeuf, C.M., Isaacs, S.J., Tran, N.H., and Kim, B. 2001. Thermal effects on reverse transcription: improvement of accuracy and processivity in cDNA synthesis. Biotechniques 30: 1074.

Mallet, F., Oriol, G., Mary, C., Verrier, B., and Mandrand, B. 1995. Continuous RT-PCR using AMV-RT and Taq DNA polymerase: characterization and comparison to uncoupled procedures. Biotechniques 18: 678-687.

Max, N., Willhauck, M., Wolf, K., Thilo, F., Reinhold, U., Pawlita, M.,

Thiel, E., and Keilholz, U. 2001. Reliability of PCR-based detection of occult tumour cells: lessons from real-time RT-PCR. Melanoma Res. 11: 371-378.

McDowell, D.G., Burns, N.A., and Parkes, H.C. 1998. Localised sequence regions possessing high melting temperatures prevent the amplification of a DNA mimic in competitive PCR. Nucleic Acids Res. 26: 3340-3347.

Meijerink, J., Mandigers, C., van de, L.L., Tonnissen, E., Goodsaid, F., and Raemaekers, J. 2001. A novel method to compensate for different amplification efficiencies between patient DNA samples in quantitative real-time PCR. J. Mol. Diagn. 3: 55-61.

Mizuno, T., Nagamura, H., Iwamoto, K.S., Ito, T., Fukuhara, T., Tokunaga, M., Tokuoka, S., Mabuchi, K., and Seyama, T. 1998. RNA from decades-old archival tissue blocks for retrospective studies. Diagn. Mol. Pathol. 7: 202-208.

Mizuno, Y., Carninci, P., Okazaki, Y., Tateno, M., Kawai, J., Amanuma, H., Muramatsu, M., and Hayashizaki, Y. 1999. Increased specificity of reverse transcription priming by trehalose and oligo-blockers allows high-efficiency window separation of mRNA display. Nucleic Acids Res. 27: 1345-1349.

Myers, T.W. and Gelfand, D.H. 1991. Reverse transcription and DNA amplification by a *Thermus thermophilus* DNA polymerase. Biochemistry 30: 7661-7666.

Nedelman, J., Heagerty, P., and Lawrence, C. 1992. Quantitative PCR with internal controls. Comput. Appl. Biosci. 8: 65-70.

Peccoud, J. and Jacob, C. 1996. Theoretical uncertainty of measurements using quantitative polymerase chain reaction. Biophys. J. 71: 101-108.

Pennington, J., Garner, S.F., Sutherland, J., and Williamson, L.M. 2001. Residual subset population analysis in WBC-reduced blood components using real-time PCR quantitation of specific mRNA. Transfusion. 41: 1591-1600.

Pfaffl, M.W. 2001. A new mathematical model for relative quantification in real-time RT-PCR. Nucleic Acids Res. 29: E45.

Pfaffl, M.W., Horgan, G.W., and Dempfle, L. 2002. Relative expression software tool (REST) for group-wise comparison and statistical analysis of relative expression results in real-time PCR. Nucleic Acids Res. 30: e36.

Pikaart, M.J. and Villeponteau, B. 1993. Suppression of PCR amplification by high levels of RNA. Biotechniques. 14: 24-25.

Poddar, S.K., Sawyer, M.H., and Connor, J.D. 1998. Effect of inhibitors in clinical specimens on Taq and Tth DNA polymerase-based PCR amplification of influenza A virus. J. Med. Microbiol. 47: 1131-1135.

Pomp, D. and Medrano, J.F. 1991. Organic solvents as facilitators of polymerase chain reaction. Biotechniques. 10: 58-59.

Raaijmakers, M.H., van Emst, L., De Witte, T., Mensink, E., and Raymakers, R.A. 2002. Quantitative assessment of gene expression in highly purified hematopoietic cells using real-time reverse transcriptase polymerase chain reaction. Exp. Hematol. 30: 481-487.

Raeymaekers, L. 1993. Quantitative PCR: theoretical considerations with practical implications. Anal. Biochem 214: 582-585.

Rees, W.A., Yager, T.D., Korte, J., and Von Hippel, P.H. 1993. Betaine can eliminate the base pair composition dependence of DNA melting. Biochemistry. 32: 137-144.

Reischl,U. and B.Kochanowski, 1999, Quantitative PCR, in: U.Reischl and B.Kochanowski, eds., Quantitative PCR protocols (Humana Press Inc, Totowa, NJ) 3-30.

Ricchetti, M. and Buc, H. 1990. Reverse transcriptases and genomic variability: the accuracy of DNA replication is enzyme specific and sequence dependent. EMBO J. 9: 1583-1593.

Ross, J. 1995. mRNA stability in mammalian cells. Microbiol. Rev. 59: 423-450.

Santoro, M.M., Liu, Y., Khan, S.M., Hou, L.X., and Bolen, D.W. 1992. Increased thermal stability of proteins in the presence of naturally occurring osmolytes. Biochemistry 31: 5278-5283.

Sarkar, G., Kapelner, S., and Sommer, S.S. 1990. Formamide can dramatically improve the specificity of PCR. Nucleic Acids Res. 18: 7465.

Schmittgen, T.D. and Zakrajsek, B.A. 2000. Effect of experimental treatment on housekeeping gene expression: validation by real-time, quantitative RT-PCR. J. Biochem. Biophys. Methods 46: 69-81.

Selvey, S., Thompson, E.W., Matthaei, K., Lea, R.A., Irving, M.G., and Griffiths, L.R. 2001. Beta-actin an unsuitable internal control for RT-PCR. Mol. Cell. Probes 15: 307-311.

Shammas, F.V., Heikkila, R., and Osland, A. 2001a. Fluorescence-based method for measuring and determining the mechanisms of recombination in quantitative PCR. Clin Chim. Acta 304: 19-28.

Shammas, F.V., Heikkila, R., and Osland, A. 2001b. Improvement of quantitative PCR reproducibility by betaine as determined by fluorescence-based method. Biotechniques 30: 950-2, 954.

Shanmugam, V., Sell, K.W., and Saha, B.K. 1993. Mistyping ACE heterozygotes. PCR Methods Appl. 3: 120-121.

Shimomaye, E. and Salvato, M. 1989. Use of avian myeloblastosis virus reverse transcriptase at high temperature for sequence analysis of highly structured RNA. Gene Anal. Tech. 6: 25-28.

Sidhu, M.K., Liao, M.J., and Rashidbaigi, A. 1996. Dimethyl sulfoxide improves RNA amplification. Biotechniques 21: 44-47.

Siebert, P.D. and Larrick, J.W. 1992. Competitive PCR. Nature. 359: 557-558.

Simpson, D., Crosby, R.M., and Skopek, T.R. 1988. A method for specific cloning and sequencing of human hprt cDNA for mutation analysis. Biochem. Biophys. Res. Commun. 151: 487-492.

Sirover, M.A. 1999. New insights into an old protein: the functional diversity of mammalian glyceraldehyde-3-phosphate dehydrogenase. Biochim. Biophys. Acta 1432: 159-184.

Smieja, M., Mahony, J.B., Goldsmith, C.H., Chong, S., Petrich, A., and Chernesky, M. 2001. Replicate PCR testing and probit analysis for detection and quantitation of *Chlamydia pneumoniae* in clinical specimens. J. Clin. Microbiol. 39: 1796-1801.

Soong, R., Beyser, K., Basten, O., Kalbe, A., Rueschoff, J., and Tabiti, K. 2001. Quantitative reverse transcription-polymerase chain reaction detection of cytokeratin 20 in noncolorectal lymph nodes. Clin.Cancer Res. 7: 3423-3429.

Souaze, F., Ntodou-Thome, A., Tran, C.Y., Rostene, W., and Forgez, P. 1996. Quantitative RT-PCR: limits and accuracy. Biotechniques 21: 280-285.

Spanakis, E. 1993. Problems related to the interpretation of autoradiographic data on gene expression using common constitutive transcripts as controls. Nucleic Acids Res. 21: 3809-3819.

Specht, K., Richter, T., Muller, U., Walch, A., Werner, M., and Hofler, H. 2001. Quantitative gene expression analysis in microdissected archival formalin-fixed and paraffin-embedded tumor tissue. Am. J. Pathol. 158: 419-429.

Spiess, A.N. and Ivell, R. 2002. A highly efficient method for long-chain cDNA synthesis using trehalose and betaine. Anal. Biochem. 301: 168-174.

Stanta, G. and Bonin, S. 1998a. RNA quantitative analysis from fixed and paraffin-embedded tissues: membrane hybridization and capillary electrophoresis. Biotechniques 24: 271-276.

Stanta, G., Bonin, S., and Utrera, R. 1998b. RNA quantitative analysis from fixed and paraffin-embedded tissues. Methods Mol. Biol. 86: 113-119.

Thellin, O., Zorzi, W., Lakaye, B., De Borman, B., Coumans, B., Hennen, G., Grisar, T., Igout, A., and Heinen, E. 1999. Housekeeping genes as internal standards: use and limits. J. Biotechnol. 75: 291-295.

Tricarico, C., Pinzani, P., Bianchi, S., Paglierani, M., Distante, V., Pazzagli, M., Bustin, S.A., and Orlando, C. 2002. Quantitative real-time reverse transcription polymerase chain reaction: normalization to rRNA or single housekeeping genes is inappropriate for human tissue biopsies. Anal. Biochem. 309: 293-300.

Vandesompele, J., De Paepe, A., and Speleman, F. 2002a. Elimination of primer-dimer artifacts and genomic coamplification using a two-step SYBR green I real-time RT-PCR. Anal. Biochem. 303: 95-98.

Vandesompele, J., De Preter, K., Pattyn, F., Poppe, B., Van Roy, N., De Paepe, A., and Speleman, F. 2002b. Accurate normalization of real-time geometric averaging of multiple internal control genes. Genome Biol. 3: RESEARCH0034.1

Varadaraj, K. and Skinner, D.M. 1994. Denaturants or cosolvents improve the specificity of PCR amplification of a G + C-rich DNA using genetically engineered DNA polymerases. Gene 140: 1-5.

Vlems, F., Soong, R., Diepstra, H., Punt, C., Wobbes, T., Tabiti, K., and Van Muijen, G. 2002. Effect of Blood Sample Handling and Reverse Transcriptase-Polymerase Chain Reaction Assay Sensitivity on Detection of CK20 Expression in Healthy Donor Blood. Diagn. Mol. Pathol. 11: 90-97.

Vrieling, H., Simons, J.W., and van Zeeland, A.A. 1988. Nucleotide sequence determination of point mutations at the mouse HPRT locus using *in vitro* amplification of HPRT mRNA sequences. Mutat. Res. 198: 107-113.

Wall S.J. and Edwards D.R. 2002. Quantitative Reverse Transcription-Polymerase Chain Reaction (RT-PCR): A Comparison of Primer-Dropping, Competitive, and Real-Time RT-PCRs. Anal. Biochem. 300: 269-273.

Wang, R.F., Cao, W.W., and Johnson, M.G. 1992. A simplified, single tube, single buffer system for RNA-PCR. Biotechniques 12: 702, 704.

Weisinger, G., Gavish, M., Mazurika, C., and Zinder, O. 1999. Transcription of actin, cyclophilin and glyceraldehyde phosphate dehydrogenase genes: tissue- and treatment-specificity. Biochim. Biophys. Acta 1446: 225-232.

Weissensteiner, T. and Lanchbury, J.S. 1996. Strategy for controlling preferential amplification and avoiding false negatives in PCR typing. Biotechniques 21: 1102-1108.

Wiedbrauk, D.L., Werner, J.C., and Drevon, A.M. 1995. Inhibition of PCR by aqueous and vitreous fluids. J. Clin. Microbiol. 33: 2643-2646.

Willey, J.C., Crawford, E.L., Jackson, C.M., Weaver, D.A., Hoban, J.C., Khuder, S.A., and DeMuth, J.P. 1998. Expression measurement of many genes simultaneously by quantitative RT-PCR using standardized mixtures of competitive templates. Am. J. Resp. Cell. Mol. Biol. 19: 6-17.

Wilson, I.G. 1997. Inhibition and facilitation of nucleic acid amplification. Appl. Environ. Microbiol. 63: 3741-3751.

Winer, J., Jung, C.K., Shackel, I., and Williams, P.M. 1999. Development and validation of real-time quantitative reverse transcriptase-polymerase chain reaction for monitoring gene expression in cardiac myocytes *in vitro*. Anal. Biochem. 270: 41-49.

Winship, P.R. 1989. An improved method for directly sequencing PCR amplified material using dimethyl sulphoxide. Nucleic Acids Res. 17: 1266.

Wu, D.Y., Ugozzoli, L., Pal, B.K., Qian, J., and Wallace, R.B. 1991. The effect of temperature and oligonucleotide primer length on the specificity and efficiency of amplification by the polymerase chain reaction. DNA Cell Biol. 10: 233-238.

Zhang, J. and Byrne, C.D. 1997a. A novel highly reproducible quantitative competitve RT PCR system. J. Mol. Biol. 274: 338-352.

Zhang, J. and Byrne, C.D. 1999. Differential priming of RNA templates during cDNA synthesis markedly affects both accuracy and reproducibility of quantitative competitive reverse-transcriptase PCR. Biochem. J. 337: 231-241.

Zhang, J., Desai, M., Ozanne, S.E., Doherty, C., Hales, C.N., and Byrne, C.D. 1997b. Two variants of quantitative reverse transcriptase PCR used to show differential expression of alpha-, beta- and gamma-fibrinogen genes in rat liver lobes. Biochem. J. 321: 769-775.

Zhong, H. and Simons, J.W. 1999. Direct comparison of GAPDH, beta-actin, cyclophilin, and 28S rRNA as internal standards for quantifying RNA levels under hypoxia. Biochem. Biophys. Res. Commun. 259: 523-526.

8

Mutation Detection by Real-Time PCR

K.J. Edwards and J.M.J Logan

Abstract

Real-time PCR is ideally suited for analysis of single nucleotide polymorphisms (SNPs) and has been increasingly used for this purpose since the advent of real-time PCR and as whole genome sequences have become available. It requires methods that are rapid, sensitive, specific and inexpensive, and several real-time methods have evolved which fulfil these requirements. Additionally real-time PCR is a technique that is readily amenable to automation and no post-PCR handling is required. Different formats have been applied including hybridisation probes with melting curve analysis, hydrolysis probes, molecular beacons and scorpion primers. SNP detection by real-time PCR has found applications in diagnosis of human disease, pharmacogenetics, clinical microbiology and drug development, and has replaced techniques such as sequencing, single strand conformation polymorphism and restriction enzyme digestion.

Principles of Mutation Detection

The detection of mutations is a fast growing field of increasing importance for many areas of science including diagnosis of human disease, pharmacogenetics, drug development and microbiology. Mutations are classed as one or more changes in the DNA bases and can include large re-arrangements such as translocations, inversions and gene insertions/deletions or small alterations such as point mutations and base insertions/deletions (SNPs). SNPs are the commonest type of DNA sequence variations and can occur once every 100-300 bases. As the requirement for rapid, reliable, sensitive and inexpensive methods for SNP detection grow the number of techniques available also increases, each with inherent strengths and weaknesses. Most of these techniques can be divided into hybridisation-based or enzyme-based methods and have been extensively reviewed (Syvanen, 2001; Kirk *et al.*, 2002; Kwok, 2002).

Real-time PCR, a hybridisation-based method, has become widely used for mutation detection. The systems are flexible with a number of different probe systems that can be used and there is additional flexibility in the design of the probes. Several formats have evolved including, hybridisation probes, hydrolysis probes, molecular beacons and scorpion primers. These methods are sensitive and specific, inexpensive, rapid (some assays can be performed in as little as 30 minutes) and they are both easy to perform and interpret the results. All of the available platforms are semi-automated and none require additional post-PCR handling for example, agarose gel analysis. Depending on the platform used real-time PCR is suitable for low to medium sample throughput. The greatest advantage of these systems is the quality of the data that is generated, known mutations are easily detected and there are possibilities with some of the systems to detect new mutations. Due to increased demand there is now a number of commercially available tests developed, especially for the Applied Biosystems Sequence Detection Systems (ABI 7700 and 7000) and the LightCycler.

In this chapter some of the probe formats will be critically discussed in more detail, as well as exploring some examples where mutation detection has been widely used.

Hybridisation Probes

Mutation detection using hybridisation probes can be performed on the LightCycler system (Roche Molecular Diagnostics) using fluorescent resonance energy transfer (FRET) for detection (Wittwer *et al.*, 1997). In the reaction two probes are used and FRET occurs when the donor probe is excited photometrically and transfers it energy to the acceptor probe causing the acceptor probe to emit light at a longer wavelength, which is measured by the instrument. The energy transfer can only occur when the two probes have hybridised to the target and are in close proximity.

Hybridisation probes can be used for mutation detection by performing a melting curve analysis directly after completing the amplification cycles. As the temperature is increased, the probes 'melt' or dissociate from the target at specific temperatures. The melting temperature (Tm) is the temperature at which 50% of the probe is dissociated from the template and this is observed as a decrease in fluorescent signal. When a mutation occurs in the target sequence the probe melts at a lower temperature than that of a perfect match (Lohmann *et al.*, 2000). The melting temperatures are displayed as melting peaks by the LightCycler software and this allows wild-type, mutant and heterozygous genotypes to be distinguished (Lay and Wittwer 1997). Figure 1 shows an example of a typical melting curve.

Probe Design

The detection probe covers the mutation of interest and is usually 15-30 bases in length. The anchor probe requires a higher Tm than the detection probe or primers, typically 5-10°C higher. This will cause the detection probe to melt first and consequently the decrease in fluorescence is due to the detection probe, as a result the presence or

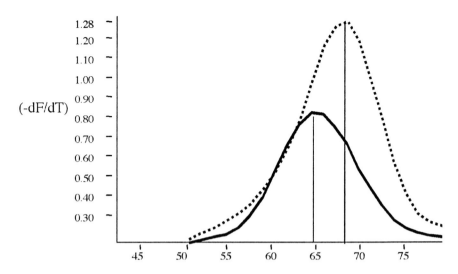

Figure 1. Melting curve analysis of wild type (dashed line) and mutant sequence (solid line) obtained using hybridisation probes on the LightCycler. The sequence containing the mutation had a Tm 4°C lower than the perfectly matched sequence and can easily be distinguished.

absence of the mutation will be observed. The melting of the anchor probe is not detected. The higher Tm of this probe can be due to G-C content or probe length and as for any hybridisation probe the optimum spacing between the two probes is 1-3 bases. A useful tool for accurately predicting the melting temperature of the hybridisation probe is the Tm tool software available from Idaho Technology (http://www.idahotec.com) (Bernard *et al.*, 1998).

The detection probes can be designed to either the mutant or wild-type sequence and to either the sense or anti-sense strand. Consideration should be given to the probe sequence that causes the most destabilised hybrids if the target contains a mismatch, in other words the hybrid containing the mutation will have a much lower Tm than the perfect match. Each mismatch has a different effect on the stability, for example, G-C mismatches are the most stable and consequently result in the smallest Tm shift, whereas T-C mismatches are the least stable and cause the largest Tm shift (Peyret *et al.*, 1999). However, even the most stable mismatches can adequately separate mutants from wild-type (Bernard *et al.*, 1998; Blomeke and Shields 1999).

PCR Conditions

The melting curve analysis is performed immediately after amplification by briefly denaturing the PCR product followed by cooling to 5-10°C below the Tm of the detection probe. The temperature is then slowly raised at a transition rate of 0.1-0.2°C per second whilst continuously monitoring fluorescence. As the probe melts the decrease in fluorescent signal is observed as a melting curve which is subsequently converted to a melting peak by the LightCycler software using the negative derivative of the melting curve. The melting peaks are easier to visualise and allow accurate determination of the Tm. To achieve good melting curves it is sometimes necessary to use asymmetric PCR to increase the amount of the target strand specific for the probe (Lyon *et al.*, 1998; Phillips *et al.*, 2000). This improves the melting curve because increased amplification of the target strand can reduce the competition between the template strands and template-probe hybridisations.

It is possible to detect more than one mutation during a reaction either by being able to easily resolve different melting peaks by Tm or by using different colour dyes for detection. Multiplexing by Tm uses the same acceptor probe with two different detector probes. The detector probes can be designed with different Tms by increasing the length of one or by differences in G-C content. It is possible to differentiate four melting peaks *i.e.* a perfect match and a mismatch for two mutations in this way (Bernard *et al.*, 1998). Alternatively there are two detection channels on the LightCycler, which can be used to detect LC Red 640 and LC Red 705. Due to spectral overlap between the channels on the LightCycler it is necessary to correct for any cross talk between the dyes by running a colour compensation experiment on the machine. This calibration run monitors the emission from the different dyes over a temperature gradient and can then be applied to subsequent runs to remove the cross talk. Multiplexing the reactions significantly increases the information that can be obtained whilst minimising time and reagent costs. As for any multiplex, changes to cycling conditions and reaction conditions especially primer concentration and magnesium concentration may need to be made in order for all reactions to run optimally.

Advantages/Disadvantages

Hybridisation probes and melting curve analysis on the LightCycler offer a very rapid system for detection of mutations. A mutation detection assay can be completed in as little as 30 minutes, with the melting curve analysis requiring only five minutes. The system is semi-automated and requires no post-PCR handling, reducing the risk of PCR contamination. The results are easy to interpret and it is possible to detect previously unknown mutations provided they occur within the probe-binding site. Optimal probes are relatively simple to design. Standard primer design software can be used to calculate the binding efficiency and Tm, with flexibility in which DNA strand is targeted in order to design an optimal probe. The disadvantages of this system are the potential for multiplexing and the number of samples, which can be processed. The reactions can be multiplexed by Tm, which is dependent on designing probes with sufficiently different Tms that will be resolved by the software. In practice the melting peaks do not always discriminate as expected. The reactions can also be multiplexed by fluorophore, at present on the LightCycler only two dyes can be used although this has been increased to four with the release of the version 2.0 instrument. On the LightCycler the number of samples is limited to 32 although the PCRs can be performed quickly, so there is potential to perform multiple runs. Hybridisation probes can also be run on other platforms including the iCycler (BioRad) and the Rotor-Gene (Corbett Research).

ResonSense Probes

ResonSense probes are similar to hybridisation probes except that only a single probe is used (Lee *et al.*, 2002). The probe is labelled with a fluorescent dye and FRET occurs between the probe and an intercalating dye, such as SYBR Green I present in solution. As with hybridisation probes ResonSense probes can be used for mutation detection by performing melting curve analysis (Gibson *et al.*, 1999; Logan *et al.*, 2001; Edwards *et al.*, 2001). For design the same considerations as for hybridisation probes are required. As with hybridisation probes the probe requires to be blocked at the 3' end to prevent it from acting as a primer, and this can be achieved either by placing the fluorescent label at the 3' end or by adding a blocker such as biotin or phosphate.

These probes offer an advantage over conventional hybridisation probes where sequence constraints prevent design of two probes. It may also be simpler to optimise the reactions with a single probe (Lee *et al.*, 2002).

Hydrolysis Probes

An alternative hybridisation technique uses hydrolysis probes to detect PCR products in real-time (Livak *et al.*, 1995). Hydrolysis probes are labelled at the 5' end with a fluorescent reporter molecule and at the 3' end with a quencher molecule. During extension the 5 '→ 3' exonuclease activity of *Taq* polymerase cleaves the probe, emitting signal due to the separation of the reporter from the quencher. For mutation detection two probes are required, one targeting the wild-type sequence and the other targeting the mismatched sequence. These probes are labelled with different fluorescent reporter molecules to differentiate amplification of each of the sequences. Mismatches between the probe and the target reduce the efficiency of probe hybridisation and furthermore *Taq* polymerase is more likely to displace a mismatched probe rather than cleaving it, with no release of the reporter dye.

Design of Hydrolysis Probes

Standard design features that apply to other chemistries also apply to hydrolysis probes including length of 20-40 nucleotides, a G-C content of 40-60%, no runs of a particular nucleotide especially G, no repeated sequence motifs and a Tm of at least 5°C higher than the primers to ensure the probe has bound to the template before extension of the primers can occur (Landt, 2001). Two reporter dyes can be used to label hydrolysis probes for mutation detection, namely VIC and FAM and typically these probes are quenched by TAMRA. Mismatches near the end of probes are not disruptive to hybridisation so it is recommended that the mutation is near the centre of the probe (Livak *et al.*, 1995, 1995). The 5' end of the probe cannot be a guanosine residue, as a guanosine adjacent to the reporter dye will quench the reporter fluorescence even after probe cleavage (Livak *et al.*, 1995). The estimated Tm for the probes should be between 65-67°C and it is recommended that the dedicated Primer Express software (Applied Biosystems) is used for the design.

Minor-Groove Binding Probes

A recent development based on hydrolysis probes is minor-groove binding probes (MGB). In this system the TAMRA quencher dye is replaced with a non-fluorescent quencher and a minor-groove binder stabilises the probe/target molecule by folding into the minor groove of the dsDNA (Kutyavin *et al.*, 1997 and 2000). The increased stability and the increased Tm, due to the MGB, allows very short probes to be designed, typically 14 bp which is useful when sequence constraints limit the design of the probe. Additionally, because a non-fluorescent quencher is used the entire fluorescent signal observed is due to the reporter molecule which increases the accuracy of mutation detection. Software to aid with the design of MGB probes is available from Amersham Biosciences.

Advantages/Disadvantages

It is not possible to perform melting curve analysis when using hydrolysis probes as the probes are destroyed during the PCR reaction, therefore differently labelled hydrolysis probes are used for mutation detection. On most of the platforms available it is possible to mutiplex two of these probes allowing mismatched and wild-type sequence to be detected in one reaction. Hydrolysis probes were initially designed for use on the ABI Sequence Detection Systems, where two dyes can be multiplexed and 96-384 samples analysed in 2 hours. This allows a rapid and relatively inexpensive system to be used for large-scale screening. Hydrolysis probes can be run on most other real-time platforms but cannot always be multiplexed. This system is dependent on the inefficiency of the probes to bind to mismatched sequences and the fact that *Taq* polymerase is more likely to displace rather than cleave a mismatched probe. It is therefore conceivable to believe that this system can lead to false results being generated.

Molecular Beacons

Molecular Beacons are hybridising probes with a stem and loop structure (Tygai *et al.*, 1998). They are composed of sequence complementary to the target sequence that forms the loop, flanked by arm sequences that are complementary to each other and form the stem. At the end of one arm-sequence a fluorescent molecule is attached and to the end of the other arm sequence is a quenching molecule. When the molecular beacon is free in solution the stem formation keeps the two dyes in close proximity causing quenching of the fluorescent molecule. However, when the probe is bound to the target molecule the fluorescent molecule is no longer quenched leading to increased fluorescence, which is measured. Molecular beacons are ideally suited for mutation detection due to the presence of the hairpin stem which means these probes bind only to perfectly matched sequences. This significantly increases the specificity when compared to linear probes such as hydrolysis or hybridisation probes (Tygai *et al.*, 1998). The donor and acceptor dyes are not required to have overlapping spectra because they are brought into close proximity due to the confirmation of the stem and loop structure. This increases the number of potential fluorophores that can be used and detection is only limited by the hardware being used for the real-time detection (Tygai *et al.*, 1998). As with MGB probes the quencher used is a non-fluorescent quencher (DABCYL) and so all fluorescent signal is due to the specific signal. However, great attention must be given to the design of these probes because the function of them is dependent upon correct hybridisation of the stem.

Probe Design

For mutation detection the molecular beacons are designed to be complementary to the mismatched base. The loop region is designed first and contains between 18-25 bases that give a melting temperature slightly higher than the annealing temperature of the primers. This causes the probe-target hybrid to be stable during annealing when the signal detection takes place (Mhlanga and Malmberg 2001). The arm sequences are usually 5-7 nucleotides and are designed to allow the stem to dissociate 7-10°C above the detection temperature, which is usually the annealing temperature of *Taq* polymerase in PCR. The Tm of the loop structure can be calculated using most primer design software that use % GC rule whereas the Tm of the stem sequence is best calculated using a DNA folding programme which will also ensure the dominant folding pattern is the intended hairpin (Mhlanga and Malmberg 2001). The SNP should be as close as possible to the centre of the probe. A software package that designs molecular beacons as well as primers is available from Premier Biosoft (Palo Alto, CA). The primers used are recommended to be sited 20-30 bp 3' or 5' of the molecular beacon respectively. The maximum product length should be no more than 250 bp as the signal has been shown to diminish in larger amplicons possibly due to the molecular beacon being unable to invade the double strands during the PCR annealing step (Mhlanga and Malmberg 2001).

Advantages/Disadvantages

The specificity of probe binding has been demonstrated to improve considerably when a hairpin stem is used and this is particularly useful for SNP detection (Tyagi *et al.*, 1998). When a mismatch is present the probe/target hybrids are considerably less stable and dissociate at temperatures much lower than perfectly matched sequences and this temperature difference is much greater than that observed with linear probes. This allows use of one common temperature to detect several mutations (Gisendorf *et al.*, 1998). Another advantage of the hairpin stem is that it brings the label moieties close together, allowing efficient quenching. This means a larger range of fluorescent dyes can be employed offering greater flexibility for multiplexing. The main drawback of molecular beacons is the complex design ensuring that a hairpin structure will be formed. As with other probe systems molecular beacons are ideally suited to medium-scale semi-automated mutation detection.

Scorpion Primers

The previously described systems all require a fluorescent-labelled probe or probes for detection whereas the scorpion system does not (Thelwell *et al.*, 2000). Like molecular beacons scorpion primers also take advantage of the stem-loop configuration. They consist of a hairpin linked to the 5' end of a PCR primer via a "PCR stopper" that prevents read-through by *Taq* polymerase and a loop, complementary to the target sequence. Attached to the 5' end of the stem is a donor fluorophore and positioned 3' of the donor just after the loop sequence is a quencher molecule (normally methyl red), so that when the stem forms the donor and quencher are opposite. When added to the reaction the scorpion primer is incorporated into the PCR product and during the cooling the scorpion primer folds back and hybridises to the complementary sequence in the newly synthesised part of the same DNA. This eliminates quenching and an increase in fluorescence is observed (Thelwell *et al.*, 2000). The hybridisation reaction occurs within the same strand and hence occurs faster than hybridisation probes or hydrolysis probes (Didenko, 2001). Scorpion primers can be used for mutation detection by monitoring fluorescence at a temperature where the probe has dissociated from a target with a mismatch but remains bound to a complementary target (Thelwell *et al.*, 2000).

Scorpions give a strong signal and act very quickly making them suitable for high throughput mutation detection assays. They are extremely versatile and, for example, are able to discriminate the most stable mismatches by decreasing the probe length until the mismatch becomes significant. This is not possible in other systems as shortening the probe length will most likely decrease the specificity whereas the probing mechanism used by scorpions increases the specificity as the primer places the probe close to the desired binding site (Thelwell *et al.*, 2000). Scorpion primers are suitable for use on most real-time platforms however the ability to multiplex is limited by the number of available detection channels.

Advantages/Disadvantages

The scorpion primer system utilises labelled primers that form a hairpin structure similar to molecular beacons and do not require an additional labelled probe(s) to be included. This has obvious cost advantages but also allows the reactions to be performed very quickly, as it is not necessary to wait for the probe to hybridise or for hydrolysis to occur (Thelwell *et al.*, 2000). Additionally, there are no design constraints as with other systems on the position of the probe relative to the primers, which makes this an ideal system when there are sequence constraints. It is possible to multiplex scorpion primers and the only constraint is the number of dyes detectable by particular platforms. For example on the LightCycler only two dyes can be multiplexed however scorpions will also work on the ABI 7700 or the iCycler where up to four dyes can be used. The scorpion primers are ideally suited for a semi-automated system that is cost-effective and efficient at detecting multiple mutations.

Fluorescent DNA Dyes

SYBR Green I is a double-stranded specific DNA dye which can be used in real-time PCR for detection of product amplification and for detection of product Tm (Wittwer *et al.*, 1997; Ririe *et al.*, 1997). It has been used for detection of SNPs in products up to 200 bp by performing melting curve analysis (Lipsky *et al.*, 2001; Elenitoba *et al.*, 2001) although additional steps may have to be performed such as amplicon purification or addition of high concentration of SYBR Green I (Lipsky *et al.*, 2001). Another intercalating dye which has been designed specifically for mutation detection is LCGreen I (Idaho technology) (Wittwer *et al.*, 2003). Wittwer *et al.* (2003) have developed this dye and compared it to four other intercalating dyes including SYBR Green I and ethidium bromide and report that it is useful for detecting mutations without inhibiting or adversely affecting the PCR. Unlike fluorescent labelled probes, using fluorescent dyes can detect unknown mutations. When using a probe based system the probe requires to be designed specifically to the sequence of interest and although they can detect unknown mutations these must occur during the probe-binding site. Intercalating dyes can detect any mutation throughout the amplicon and can therefore be used for mutation screening with the added advantage of lower cost (Wittwer *et al.*, 2003).

Applications of Real-Time PCR for Mutation Detection

As the ability to generate huge amounts of sequence data develops, information on sequence changes increases and this can be exploited for many purposes such as clinical diagnostics and microbiology. Molecular methods for SNP detection have been available for some time, however most of these methods are time consuming and cumbersome to perform. The development of real-time PCR has led to an explosion in cheap and rapid mutation detection methods. The applications in disease diagnostics for Factor V Leiden and Cystic Fibrosis will be explored, as well as the detection of antibiotic resistance mutations in bacteria.

Factor V Leiden

The first published assay for mutation detection was the detection of the Factor V Leiden mutation (Lay and Wittwer, 1997). It is a single point mutation in the factor V gene where a G is substituted for an A at position 1691, and it is a risk factor for venous thrombosis. Patients that are both homozygous and heterozygous for this mutation have an increased risk in developing the disease and, due to the high prevalence of the mutation there is an increased demand for clinical laboratories to provide fast, reliable and inexpensive tests.

The assay developed by Lay and Wittwer (1997) utilised a Cy5-labelled primer and a fluorescein labelled probe in a hybridisation system where the factor V Leiden mutation occurred 8 bases from the 5' end of the 23 bp probe. Amplification and melting curve analysis was performed on the LightCycler where after 45 rapid amplification cycles the PCR product was melted by slowly heating from 50°C to 75°C and fluorescence was continuously monitored. When wild-type sequence was detected the melting peak had a Tm of 66°C whereas a homozygous mutant resulted in a melting peak 8°C lower at 58°C. Heterozygous samples containing both sequences could also be detected and in these both melting peaks were observed. The assay was tested with 100 clinical samples and compared to the traditional method of PCR followed by restriction enzyme digestion and size separation. All 100 samples were correctly genotyped (Lay and Wittwer, 1997). The traditional method required at least 4-6 hours for completion whereas the LightCycler assay can be completed in just 30 minutes.

Assays have been developed for the ABI TaqMan using a TET-labelled and a FAM-labelled hydrolysis probe, one designed to match the wild-type sequence and the other to the mutant sequence (Sevall 2000) and for the Rotor-Gene using hybridisation probes (Ameziane et al., 2003). The latter two assays offer a higher throughput capacity than the LightCycler. A multiplex assay has also been developed to detect the factor V Leiden mutation as well as another mutation in the prothrombin gene that is associated with hereditary thrombophilia using hybridisation probes on the LightCycler (van den Bergh et al., 2000).

Cystic Fibrosis

Cystic Fibrosis (CF) is a common autosomal recessive disorder with a frequency of 0.05% and a carrier frequency of 5% (Rommens et al., 1989). It is caused by mutations in a gene on chromosome 7 (ABCC7) that produces the cystic fibrosis transmembrane conductance regulator protein. The transport of chloride ions across the membrane of epithelial cells is affected by the mutations and leads to a number of clinical symptoms. The most common mutation is a 3 bp deletion at codon 508

(F508del) and half of CF sufferers are homozygous for this mutation. A hybridisation probe assay was developed for the LightCycler that allows detection of this mutation in less than 30 minutes (Gundry *et al.*, 1999). Two fluorescein-labelled hybridisation probes were designed to match the wild-type sequence perfectly from either codon 502 to 513 or from codon 504 to codon 511. The difference in the Tm between the wild-type and sequence containing the 3 bp deletion was approximately 10°C. During detection of this F508del mutation a second mutation in the same codon was detected. The temperature difference of the second mutation (F508C) was only 5°C from the wild-type and this allowed both mutations to be detected (Gundry *et al.*, 1999). The second mutation was only observed because melting curve analysis was used and because the second mutation also occurred within the probe binding site. It would not have been detected by some methods, for example hydrolysis probes and demonstrates the potential power of melting curve analysis for mutation detection. In this case detection of the second mutation is particularly useful, as it is essential to test for this mutation to rule out false positives.

In a more recent study, scorpion primers have been used to detect five common CF-related mutations (Thelwell *et al.*, 2000). In this system primers were designed close to the mutation site and in all but one the scorpion probe sequence was attached to the primer closest to the mutation site. The scorpion system was successfully used to distinguish wild-type and mutant sequence by having a probe that detects each type and which were run in separate reactions. Heterozygous samples were shown to have approximately half the level of fluorescence of wild-type sequence (Thelwell *et al.*, 2000). Depending on the real-time platform used it is possible to multiplex the scorpion primers to provide a quick and inexpensive method for detection.

Detection of Antibiotic Resistance Mutations in Bacteria

Since 1999 there has been a number of studies performed detecting drug resistance in bacteria. These drug resistances are nearly always due to multiple mutations. These are most easily detected using melting curve analysis on the LightCycler because hybridisation probes/ResonSense probes will bind to mismatched as well as perfectly matched sequences and therefore discriminate both the mutation they were designed to detect and any additional mutations which occur in the same binding site. In this way extra information can be obtained resulting in assays that detect multiple mutations whilst retaining the other benefits of the system including speed and simplicity. Table 1 summarises some of the organisms to which the LightCycler/hybridisation probe system has been applied. Detecting drug resistance mutations in this way has led to faster more reliable methods supplying information that is increasingly important as organisms develop more and more drug resistances. Two organisms which are slow or difficult to grow, will be explored in more detail.

Mycobacteria tuberculosis

Mycobacterium tuberculosis is responsible for three million deaths a year worldwide (Raviglione *et al.*, 1995). Recommended treatment for tuberculosis (TB) comprises a combination of four drugs: rifampicin, isoniazid, pyrazinamide and ethambutol with or without streptomycin. Resistance has emerged to all of these drugs and multidrug resistance (MDR) in which the isolate is resistant to rifampicin and to at least one other drug is becoming more common (Brown *et al.*, 2000). Rapid identification of drug resistance is necessary to minimise transmission of drug-resistant strains and to allow alternative drug therapy to be started (Edlin *et al.*, 1992; Espinall *et al.*, 2001). Rifampicin resistance

Table 1. LightCycler assays for detection of antibiotic resistance mutations in bacteria.

Organism	Gene	Antibiotic	Publication
Staphylococcus aureus	*grlA*	Fluoroquinolone	Lapierre *et al.*, 2003
Neisseria gonorrhoeae	*gyrA*	Ciprofloxacin	Li *et al.* 2002
Campylobacter coli	*gyrA*	Ciprofloxacin	Carattoli *et al.*, 2002
Enterococcus faecalis and *E. faecium*	23S rRNA	Oxazolidinone	Woodford *et al.*, 2002
Salmonella enterica	*gyrA*	Cyclohexane	Liebana *et al.*, 2002

is well characterised and 95% of resistant *M. tuberculosis* strains have a mutation within an 81-bp region of the *rpoB* gene which encodes the β-subunit of the RNA polymerase (Kapur *et al.*, 1994). Resistance to isoniazid is more complex with at mutations in at least four genes having been reported, though the highest proportion of mutations is in the *katG* gene.

In one study two changes in the *rpoB* gene at codon 531 and 526 and the most common mutation in the *katG* gene was detected using hybridisation probes on the LightCycler (Torres *et al.*, 2000). The detector probe was designed to cover the mutation site. Using the rpoB probe the wild-type sequence gave a melting peak of 64°C, whereas a 2°C increase was detected when strains had a mutation at codon 531 and when strains had a mutation at codon 526 the melting peak was observed at 58°C *i.e.* a difference of 6°C. Using the katG probe, the wild-type strains melted at 72°C and two different mutations at codon 531 were detected, one melting 3°C lower and the other melting 5°C lower (Torres *et al.*, 2000). This assay allowed detection of four of the most common drug-resistance related mutations using just two hybridisation probe pairs.

In a similar study, ResonSense probes (referred to as biprobes in the publication) were used to detect four rifampicin resistance mutations (Edwards *et al.*, 2001). Three ResonSense probes were designed to cover three of the mutations and were labelled with Cy5. FRET occurred between the SYBR Green I, which was added to the PCR and between the Cy5-labelled probe and melting peaks were observed. The first probe spanned a mutation at codon 511 and resulted in a 4°C decrease in Tm when the mutation was present. The second probe spanned the mutation at 516 and resulted in a 7°C decrease in probe Tm. The final probe spanned a mutation at codon 526 and resulted in three melting peaks. The wild-type sequence melted at 61°C, whereas the mutation at codon 526 resulted in a melting temperature of 49°C and a mutation at codon 531, a mutation not spanned by the probe, resulted in a melting temperature of 66°C. It was not expected that the mutation at 531 would be detected however, it is thought that this sequence locus has the potential to form a relatively stable hairpin structure when in single-stranded form. It is believed that this structure destabilises the probe/amplicon duplex and in strains with the 531 mutation, the single-stranded structures are less able to compete with the duplex structure. This explains why a higher melting temperature was observed compared to wild-type and was reproducibly demonstrated in 27 strains tested (Edwards *et al.*, 2001). By using three ResonSense probes which all worked under the same cycling conditions, 98% of the mutations that were responsible for rifampicin resistance were easily detected. This study demonstrated that ResonSense probes are very flexible and can be used to detect mutations additional to those they were designed to detect but they can be affected by secondary structure and may not always perform as expected.

In a novel approach two hybridisation probe pairs were used to detect mutations in the *rpoB* gene but the probes were not used in the standard anchor-sensor design (Garcia de Viedema *et al.*, 2002). Instead two adjacent pairs of FRET probes were designed to cover the entire 81 bp region of the *rpoB* gene. One pair of probes that were specific for the 5' half of the region and were labelled with fluorescein and LC Red 640. The other pair were specific for the 3' region and were labelled with fluorescein and LC Red 705. This allowed the probes to be multiplexed on the LightCycler so that a single-tube approach could be used. Using this method 12 different melting peaks were observed relating to 11 different mutations within the *rpoB* gene. In this way it is possible to search for a wider number of mutations and is particularly useful for detecting emerging mutations (Garcia de Viedema *et al.*, 2002). The only potential drawback is that some of the detectable mutations may not relate to resistance.

Helicobacter pylori

Helicobacter pylori is a major cause of gastritis, peptic ulcer disease and gastric cancer (Blaser, 1997). Eradication using antibiotic therapy is recommended and one of the most widely used drugs is clarithromycin. However, resistance to clarithromycin is on the increase and is leading to treatment failure. Clarithromycin resistance is associated with three mutations in the peptidyltransferase encoding region of the 23S rRNA gene (Versalovic *et al.*, 1996; Occhialini *et al.*, 1997; Stone *et al.*, 1997).

The first assay developed to detect these mutations utilised a single ResonSense probe labelled with Cy5 and was used on the LightCycler (Gibson *et al.*, 1999). Clarithromycin sensitive strains had a probe melting peak at 68°C. Detection of two different mutations resulted in melting peaks of 63°C or 58°C. This method was originally designed for use on pure culture of the organism but has subsequently been applied to biopsy samples (Chisholm *et al.*, 2001). This significantly increases the speed of diagnosis of drug resistance and allows appropriate therapy to commence.

Conclusions

As a result of whole genome sequencing projects there has evolved a need to be able to accurately detect mutations in a cost-effective, rapid and simple approach. Real-time PCR is suited for this approach and a number of different systems, both platforms and chemistries, have been evaluated for this purpose. Traditional methods that were used included restriction enzyme digestion, sequencing, single-stranded conformation polymorphism (SSCP) and specific PCR followed by direct hybridisation. Most of these methods are time consuming, slow and can often be expensive to perform especially when a large number of mutations are to be analysed.

Other new technologies have been applied to mutation detection and some may offer advantages over real-time PCR. Microarrays have been applied to mutation detection and allow numerous sequences to be analysed in parallel (Tillib and Mirzabekov 2001). Microarrays consist of amplifying PCR products that are either attached to the surface of a slide and are interrogated with specific probes or alternatively the specific probes are attached to the slide and the PCR products hybridised to them. In this way mutations can be detected in a large-scale format. The major advantage of this method is the ability to scan for large numbers of mutations in one experiment. However, at present it is difficult to distinguish all polymorphisms and difficult to detect low-level changes (Kirk *et al.*, 2002). Microarrays are still dependent on performing PCR amplification and production of the array can be a very costly procedure. It is highly likely however, as microarray technology improves and costs are reduced that this technology will be widely used for mutation detection especially when large-scale screening is required.

Sequencing has always had a role in mutation detection because of its accuracy and robustness although it is a time consuming and expensive method. However, pyrosequencing is ideally suited for SNP detection (Alderborn *et al.*, 2000). Pyrosequencing is a novel sequencing method that measures the amount of pyrophosphate released as a result of nucleotide incorporation onto an amplified target (Ronaghi, 2003). It is a highly automated system that results in short sequence reads within 10-20 minutes for 96 samples. It is ideally suited for mutation detection where the base changes are sited close together. At present it is more expensive than using real-time PCR but it is finding an increasing role in mutation detection.

Real-time PCR offers a method for mutation detection that is extremely flexible, a large number of different approaches can be applied depending on the target of interest. The methods are all semi-automated and offer low to medium throughput depending on the platform used. The methods are very accurate in identifying the mutation and all platforms offer a close-tubed system that minimises cross-contamination in the laboratory. Additionally, the methods are relatively inexpensive and are easy to perform with simple analysis of results.

References

Alderborn, A., Kristofferson, A. and Hammerling, U. 2000. Determination of single-nucleotide polymorphisms by real-time pyrophosphate DNA sequencing. Genome Res. 10: 1249-1258.

Ameziane, N., Lamotte, M., Lamoril, J., Lebret, D., Deybach, J.C., Kaiser, T. and de Prost, J. 2003. Combined factor V leiden (G1691A) and prothrombin (G20210A) genotyping by multiplex real-time polymerase chain reaction using fluorescent resonance energy transfer hybridization probes on the Rotor-Gene 2000. Blood Coagul. Fibrinloysis. 14: 421-424.

Bernard, P.S., Lay, M.J. and Wittwer, C.T. 1998. Integrated amplification and detection of the C677T point mutation in the methylenetetrahy drofolate reductase gene by fluorescence resonance energy transfer and probe melting curves. Anal. Biochem. 255: 101-107

Blaser, M.J. 1997. Ecology of *Helicobacter pylori* in the human stomach. J. Clin. Invest. 100: 759-762.

Blomeke, B. and Shields, P.G. 1999. Laboratory methods for the determination of genetic polymorphisms in humans. IARC Sci. Publ. 148: 133-147.

Brown, T.J, Tansel, O. and French, G.L. 2000. Simultaneous identification and typing of multi-drug-resistant *Mycobacterium tuberculosis* isolates by analysis of *pncA* and *rpoB*. J. Med. Microbiol. 49: 651-656.

Carattoli, A., Dionisi, A. and Luzzi, I. 2002. Use of a LightCycler *gyrA* mutation assay for identification of ciprofloxacin-resistant *Campylobacter coli*. FEMS Microbiol. Lett. 27: 87-93.

Chisholm, S.A., Owen, R.J., Teare, E.L. and Saverymuttu, S. 2001. PCR-based diagnosis of *Helicobacter pylori* infection and real-time determination of clarithromycin resistance directly from human gastric biopsy samples. J. Clin. Microbiol. 39: 1217-1220.

Didenko, V.V. 2001. DNA probes using fluorescence resonance energy transfer (FRET): Designs and applications. BioTechniques. 31: 1106-1121.

Edlin, B.R., Tokars, J.I., Grieco, M.H., Crawford, J.T., Williams, J., Sordillo, E.M., Ong, K.R., Kilburn, J.O., Dooley, S.W. and Castro, K.G. 1992. An outbreak of multidrug-resistant tuberculosis among hospitalized patients with the acquired immunodeficiency syndrome. N. Engl. J. Med. 326: 1514-1521.

Edwards, K.J., Metherell, L.A., Yates, M and Saunders, N.A. 2001. Detection of *rpoB* mutations in *Mycobacterium tuberculosis* by biprobe analysis. J. Clin. Microbiol. 39: 3350-3352.

Elenitoba, K.S.J., Bohling, S.D., Wittwer, C.T. and King, T.C. 2001. Multiplex PCR by multicolour fluorimetry and fluorescence melting curve analysis. Nature Med. 7: 249-253.

Espinal, M.A., Laszlo, A., Simonsen, L., Boulahbal, F., Kim, S.J., Reniero, A., Hoffner, S., Rieder, H.L., Binkin, N., Dye, C., Williams, R., Raviglione, M.C. 2001. Global trends in resistance to antituberculosis drugs. World Health Organization-International Union against Tuberculosis and Lung Disease Working Group on Anti-Tuberculosis Drug Resistance Surveillance. N. Engl. J. Med. 344: 1294-1303.

Garcia de Viedma, D., del Sol Diaz Infantes, M., Lasala, F., Chaves, F., Alcala, L., Bouza, E. 2002. New real-time PCR able to detect in a single tube multiple rifampin resistance mutations and high-level isoniazid resistance mutations in *Mycobacterium tuberculosis*. J. Clin. Microbiol. 40: 988-995.

Gibson, J., Saunders, N.A., Burke, B. and Owen, R.J. 1999. Novel method for rapid determination of clarithromycin sensitivity in *Helicobacter pylori*. J. Clin. Microbiol. 37: 3746-3748.

Giesendorf, B.A.J., Vet, J.A.M., Tyagi, S., Mensik, E.J.M.G, Trijbels, F.J.M. and Blom, H.J. 1998. Molecular beacons: a new approach for semiautomated mutation analysis. Clin. Chem. 44: 482-486.

Gundry, C.N., Bernard, P.S., Herrmann, M.G., Reed, G.H. and Wittwer, C.T. 1999. Rapid F508del and F508C assay using fluorescent hybridisation probes. Genet. Test. 3: 365-370.

Kapur, V., Li, L.L., Iordanescu, S., Hamrick, M.R., Wanger, A., Kreiswirth, B.N. and Musser, J.M. 1994. Characterization by automated DNA sequencing of mutations in the gene (*rpoB*) encoding the RNA polymerase beta subunit in rifampin-resistant *Mycobacterium tuberculosis* strains from New York City and Texas. J. Clin. Microbiol. 32: 1095-1098.

Kirk, B.W., Feinsod, M., Favis, R., Kliman, R.M. and Barany, F. 2002. Single nucleotide polymorphism seeking long term association with complex disease. Nuc. Acid. Res. 30: 3295-3311.

Kutyavin, IV, Lukhtanov, EA, Gamper, HB, Meyer, RB. 1997. Oligonucleotides with conjugated dihydropyrroloindole tripeptides: base composition and backbone effects on hybridization. Nuc. Acid Res. 15: 3718-3723.

Kutyavin, I.V., Afonina, I.A., Mills, A, Gorn, V.V., Lukhtanov, E.A.,

Belousov, E.S., Singer, M.J., Walburger, D.K., Lokhov, S.G., Gall, A.A., Dempcy, R., Reed, M.W., Meyer, R.B. and Hedgpeth, J. 2000. 3'-minor groove binder-DNA probes increase sequence specificity at PCR extension temperatures. Nuc. Acid Res. 28: 655-661.

Kwok, P-Y. 2002. SNP genotyping with fluorescence polarization detection. Human Mutation. 19: 315-323.

Lapierre, P., Huletsky, A., Fortin, V., Picard, F.J., Roy, P.H., Ouellette, M. and Bergeron, M.G. 2003. Real-time PCR assay for detection of fluoroquinolone resistance associated with *grlA* mutations in *Staphylococcus aureus*. J. Clin. Microbiol. 41: 3246-3251.

Landt, O. 2001. In Meuer, S., Wittwer, C.T., Nakagawara, K (eds). Rapid Cycle Real-time PCR:Methods and applications. Springer Verlag, Germany. 35-41.

Lay, M.J. and Wittwer, C.T. 1997. Real-time fluorescence genotyping of Factor V Leiden during rapid-cycle PCR. Clin. Chem. 43: 2262-2267.

Lee, M.A., Siddle, S.L. and Hunter, R.P. 2002. ResonSense®: Simple linear probes for quantitative homogeneous rapid polymerase chain reaction. Anal. Chimic. Acta. 457: 61-70.

Li, Z., Yokoi, S., Kawamura, Y., Maeda, S., Ezaki, T and Deguchi, T. 2002. Rapid detection of quinolone resistance-associated *gyrA* mutations in *Neisseria gonorrhoeae* with a LightCycler. J. Infect. Chemother. 8: 145-150.

Liebana, E., Clouting, C., Cassar, C.A., Randall, L.P., Walker, R.A., Threlfall, E.J., Clifton-Hadley, F.A., Ridley, A.M. and Davies, R.H. 2002. Comparison of *gyrA* mutations, cyclohexane resistance, and the presence of class I integrons in *Salmonella enterica* from farm animals in England and Wales. J. Clin. Microbiol. 40: 1481-1486.

Lipsky, R.H., Mazzanti, C.M., Rudolph, J.G., Xu, K., Vyas, G., Bozak, D., Radel, M.Q. and Goldman, D. 2001. DNA melting analysis for detection of single nucleotide polymorphisms. Clin. Chem. 47: 635-644.

Livak, K.J., Flood, S.J.A., Marmaro, J., Giusti, W. and Deetz, K. 1995. Oligonucleotides with fluorescent dyes at opposite ends provide a quenched probe system useful for detecting PCR product and nucleic acid hybridization. PCR Methods Appl. 4: 357-362.

Logan, J.M., Edwards, K.J., Saunders, N.A., Stanley, J. 2001. Rapid identification of *Campylobacter* spp. by melting peak analysis of biprobes in real-time PCR. J. Clin. Microbiol. 39: 2227-2232.

Lohmann, S., Lehmann, L. and Tabiti, K. 2000. Fast and flexible single nucleotide polymorphism (SNP) detection with the LightCycler system. Biochemica. 4: 23-28.

Lyon, E., Millson, A., Phan, T and Wittwer, C.T. 1998. Detection and identification of base alterations within the region of Factor V Leiden by fluorescent melting curves. Mol. Diag. 3: 203-210.

Mhlanga, M.M. and Malmberg, L. 2001. Using molecular beacons to detect single-nucleotide polymorphisms with real-time PCR. Methods. 25: 463-471.

Occhialini, A., Urdaci, M., Doucet-Populaire, F., Bebear, C.M., Lamouliatte, H. and Megraud, F. 1997. Macrolide resistance in *Helicobacter pylori*: rapid detection of point mutations and assays of macrolide binding to ribosomes. Antimicrob. Agents. Chemother. 41: 2724-2728.

Peyret, N., Seneviratne, P.A., Allawi, H.T. and SantaLucia, J. 1999. Nearest-neighbor thermodynamics and NMR of DNA sequences with internal A.A, C.C, G.G, and T.T mismatches. Biochemistry. 23: 3468-3477.

Phillips, M., Meadows, C.A., Huang, M.Y., Millson, A. and Lyon, E. 2000. Simultaneous detection of C282Y and H63D hemochromatosis mutations by dual-colour probes. Mol. Diag. 5: 107-116.

Raviglione, M.C., Snider, D.E. and Kochi, A. 1995. Global epidemiology of tuberculosis. Morbidity and mortality of a worldwide epidemic. JAMA. 273: 220-226.

Ririe, K.M., Rasmussen, R.P. and Wittwer, C.T. 1997. Product differentiation by analysis of DNA melting curves during the polymerase chain reaction. Anal. Biochem. 245: 154-160.

Rommens, J.M., Iannuzzi, M.C., Kerem, B., Drumm, M.L., Melmer, G., Dean, M., Rozmahel, R., Cole, J.L., Kennedy, D. and Hidaka, N. 1989. Identification of the cystic fibrosis gene: chromosome walking and jumping. Science. 8: 1059-1065.

Ronaghi, M. 2003. Pyrosequencing for SNP genotyping. Methods Mol. Biol. 212: 189-195.

Sevall, S.J. 2000. Factor V Leiden genotyping using real-time fluorescent polymerase chain reaction. Mol. Cell Probes. 14: 249-253.

Stone, G.G., Shortridge, D., Versalovic, J., Beyer, J., Flamm, R.K., Graham, D.Y., Ghoneim, A.T. and Tanaka, S.K. 1997. A PCR-oligonucleotide ligation assay to determine the prevalence of 23S rRNA gene mutations in clarithromycin-resistant *Helicobacter pylori*. Antimicrob. Agents Chemother. 41: 712-714.

Syvanen, A.C. 2001. Accessing genetic variation: genotyping single nucleotide polymorphisms. Nat. Rev. Genet. 2: 930-942.

Thelwell, N., Millington, S., Solinas, A., Booth, J. and Brown, T. 2000. Mode of action and application of Scorpion primers to mutation detection. Nuc. Acid Res. 28: 3752-3761.

Tillib, S.V. and Mirzabekov, A.D. 2001. Advances in the analysis of DNA sequence variations using oligonucleotide microchip technology. Curr. Opin. Biotechnol. 12: 53-58.

Torres, M.J., Criado, A., Palomares, J.C. and Aznar, J. 2000. Use of real-time PCR and fluorimetry for rapid detection of rifampin and isoniazid resistance-associated mutations in *Mycobacterium tuberculosis*. 38: 3194-3199.

Tyagi, S., Bratu, D.P. and Kramer, F.R. 1998. Multicolour molecular beacons for allele discrimination. Nat. Biotechnol. 16: 49-53.

van den Bergh, F.A., van Oeveren-Dybicz, A.M. and Bon, M.A. 2000. Rapid single-tube genotyping of the factor V Leiden and prothrombin mutations by real-time PCR using dual-color detection. Clin. Chem. 46: 1191-1195.

Versalovic, J., Shortridge, D., Kibler, K., Griffy, M.V., Beyer, J., Flamm, R.K., Tanaka, S.K., Graham, D.Y. and Go, M.F. 1996. Mutations in 23S rRNA are associated with clarithromycin resistance in *Helicobacter pylori*. Antimicrob. Agents Chemother. 40: 477-480.

Wittwer, C.T., Ririe, K.M., Andrew, R.V., David, D.A., Gundry, R.A. and Balis, U.J. 1997. The LightCycler: a microvolume multisample fluorimeter with rapid temperature control. BioTechniques. 22: 176-181.

Wittwer, C.T., Reed, G.H., Gundry, C.N., Vandersteen, J.G. and Pyror, R.J. 2003. High-resolution genotyping by amplicon melting analysis using LCGreen. Clin. Chem. 49: 853-860.

Woodford, N., Tysall, L., Cressida, A., Stockdale, M.W., Lawson, A.J., Walker, R.A., Livermore, D.M. 2002. Detection of oxazolidinone-resistant *Enterococcus faecalis* and *Enterococcus faecium* by real-time PCR and PCR-restriction fragment length polymorphism analysis. J. Clin. Microbiol. 40: 4298-4300.

9

The Quantitative Amplification Refractory Mutation System

P. Punia and N. Saunders

Abstract

The amplification refractory mutation system (ARMS), which has also been described as allele-specific PCR (ASP) and PCR amplification of specific alleles (PASA), is a PCR-based method of detecting single base mutations (Newton *et al.*, 1989). ARMS has been applied successfully to the analysis of a wide range of polymorphisms, germ-line mutations and somatic mutations. The technique has the ability to discriminate low-levels of the mutant sequence in a high background of wild-type DNA (Billadeau *et al.*, 1991). In an ARMS PCR the terminal 3'nucleotide of one of the PCR primers coincides with the target mutation. Most applications of the method rely on 'end-point' analysis, utilising the classic gel-electrophoresis method. However, end-point analysis can only assess the presence or absence of mutant or wild-type sequences and does not give an indication of the ratio of mutant to wild-type in a mixed population of DNA. Here we describe a real-time PCR adaptation of ARMS, quantitative ARMS, that allows measurement of the size of the population of each variant in a mixture. A method for the detection of human hepatitis B virus mutations that confer resistance to the antiviral lamivudine is described as an example.

Introduction

PCR can be adapted in many ways to facilitate the analysis of DNA. Several different strategies are used for the detection of single point mutations. Current methods rely on target amplification, usually by PCR, followed by the identification of DNA variants using probes, restriction endonucleases, ligases or polymerases.

In ARMS, the primer pair is designed so that one of the 3' ends coincides with a variant nucleotide in the target sequence. When the primer mismatches the template the frequency of extension is very low and consequently the effective number of sequence copies available for amplification is greatly reduced. ARMS PCR exploits a thermostable polymerase that lacks 3' exonuclease activity (usually *Taq* polymerase). Such enzymes extend primers bound to their target sequences very inefficiently when the 3' base is mismatched. Because the 3' exonuclease activity required for mismatch repair is not present, the extension of such primers in PCR is a rare event and amplification is retarded (Figure 1). However, once mismatch extension has occurred amplification proceeds normally from the newly synthesised target strand. Thus an important principle of ARMS is that non-matching (*i.e.* non-matching at only the 3' end of one primer) template may be amplified. The factors that contribute to this apparent 'failure' of ARMS are that when very high numbers of template molecules are added to the reaction, amplification is likely to proceed because the small proportion of mismatched primer extensions reach the sensitivity threshold of the reaction. Second, certain mismatches are extended more efficiently than others so that proportion extended in each PCR cycle varies.

Detection of a single nucleotide mutation at a predetermined point in a target sequence can be achieved by running the ARMS PCR with a single primer pair and then scoring the production of amplicon as

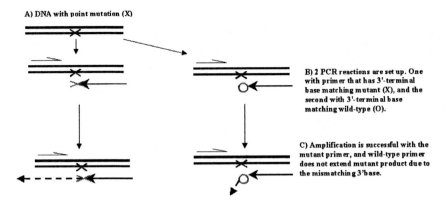

A) DNA with point mutation (X)

B) 2 PCR reactions are set up. One with primer that has 3'-terminal base matching mutant (X), and the second with 3'-terminal base matching wild-type (O).

C) Amplification is successful with the mutant primer, and wild-type primer does not extend mutant product due to the mismatching 3'base.

Figure 1. The principle of ARMS. Primers are designed to amplify wild-type and mutant sequence. The difference between these two primers is the 3' base, where one matches wild-type sequence and the other the mutant. A primer common to both reactions is present, and both reactions are run in parallel. The reaction containing the mutant primer will only extend mutant template and wild-type template will not be extended by this primer.

positive or negative. To provide a suitable control two PCR reactions are run in parallel, one with a primer matching the variant at the 3' end, and the other with the primer matching the parent sequence (Figure 1). When ARMS reactions are run in real-time instruments the difference in reaction kinetics between matching and non-matching templates contribute to large variations in the observed crossing thresholds (Cts). This has the significant advantage that it is possible to obtain clear results over a wider range of reaction conditions, than is possible when end-point analysis is used. For example, at high template concentrations a product may be produced in reactions including either set of primers. Positive and negative reactions will therefore be difficult to distinguish at the end-point but will have quite different Cts. The use of real-time instruments also allows quantitative ARMS assays to be designed. These assays can be used to determine the proportions of mutant and parent sequences present within a mixture with high sensitivity.

Efficient and sensitive methods are required to analyse gene sequence variants and quantitative ARMS therefore has many potential applications. For example, it can be used to identify genetic disease risk factors, to screen subjects for genetic susceptibility to certain environmental agents of disease and identify persons at high risk of having a child with a genetic disorder. Quantitative ARMS is particularly appropriate for monitoring the emergence of drug resistant pathogens when single nucleotide polymorphisms are appropriate markers of the resistant phenotype. The method should also be useful for the detection of residual disease in cancer. In this chapter the use of quantitative ARMS for the detection of very small quantities of lamivudine resistant human hepatitis B virus in a background of a much larger amount of sensitive virus will be used as an example.

Selection of Primers for Quantitative ARMS

In order to achieve the highest level of specificity in ARMS reactions the ideal situation would be that the mismatching primers are not extended. In reality, however, amplification of the mismatched template cannot be avoided when high levels are present in the reaction mixture. Quantitative ARMS depends upon the fact that amplification of mismatched template is less efficient than the matching PCR so that many more amplification cycles are needed to generate detectable levels of product. Primer design is important in order to maximise ARMS specificity.

Effects of the 3' Terminal Base

Previous studies on the extension of 3' base mismatched primers have all shown that the rate is dependent on the structure formed at the mismatch site. However, the different studies have not agreed on the relative rates of extension for the possible mismatch pairs. Results of five of these studies are presented in Table 1.

The discrepancies in the data obtained in these investigations can be explained by the influence of bases adjacent to the mismatched pair since these were not identical in the different studies. It should perhaps not be surprising that the local environment in which the mismatch extension is poised to occur will have a significant effect on the reaction rate. The finding that certain mismatches are extended

214

Table 1. Summary of prior studies reporting effects on PCR of mismatch between the primer 3'-nucleotide and complementary template DNA sequence.[$]

Template nucleotide	A	T	G	C
		Primer 3' nucleotide		
(Kwok et al., 1990)				
T	++++	+++	+++	+++
A	+	++++	+++	+++
C	+++	+++	++++	++
G	++	+++	++	++++
(Okayama et al., 1989)				
T	++++	-	-	++++
A	-	++++	+++	+++
C	-	+++	++++	++
G	++++	+++	++	++++
*(Huang et al., 1992)**				
T	++++	+	++	+++
A	-	++++	-	++
C	++	+	++++	-
G	-	++	-	++++
*(Day et al., 1999)***				
T	++++	+++	++++	+++
A	++	++++	+++	+++
C	++++	+++	++++	+++
G	++	+++	+++	++++
*(Ayyadevara et al., 2000)****				
T	++++	+	+++	++
A	+	++++	+	+
C	+	+	++++	++
G	+	++	+	++++

[$]The results from five papers are summarised (Okayama et al., 1989, Kwok et al., 1990, Huang et al., 1992, Day et al., 1999). If the 3' primer:template mismatch resulted in no detectable PCR amplification it is shown as -. When a PCR product was produced the relative band intensities have been converted into a 4-class scale (+ to ++++), where products generated by perfect match are assigned to ++++.

*For Huang et al. (1992) the quantitative reduction in *Taq* DNA polymerase extension of mismatched primer:template base pairing is represented as + when primer extension was reduced by 10^{-4}-10^{-5} relative to correct base pairing (++++), ++ when reduced by 10^{-3}-10^{-4}, and +++ when extension is reduced by ~10^{-2}.

**For Day et al. (1999) if product was detected by ethidium bromide staining after 10 cycles of PCR amplification, it is shown as ++++, after 20 cycles, +++; and if detectable only after 30 cycles, it is shown as ++.

***For Ayyadevara et al. (2000) quantity is given as percentage comparison of first round PCR band intensities, where: + represents 0-1% of anchor primer band intensity; ++ represents 1-2%; +++ represents 2-3% and ++++ represents 3+%.

more efficiently than others is not as helpful in assay design as it might at first appear. Clearly the 3' end of any ARMS primer is set by the particular mutation under investigation. One area of flexibility is that two different mismatch primers can be designed for each mutation *i.e.* one annealing to one strand and one to the other. The structures resulting from hybridisation of these primers to their target sequences are completely different and may therefore be expected to have different rates of mismatch extension.

Effects of Mismatched Bases Close to the 3' End

Allele specific primers can be further destabilised by including bases that mismatch the template. The effect of these mismatches increases with proximity to the 3' terminus. It has been reported (Kwok *et al.*, 1990, Newton *et al.*, 1989) that the inclusion of a mismatched base adjacent to the 3' base (the n-1 base) can increase the specificity of ARMS primers. This depends upon the idea that the n-1 mismatch will have a disproportionate effect (decrease) on priming efficiency when the terminal base is also a mismatch. We have found that some n-1 or n-2 mismatches have a large effect on priming efficiency while others appear to have little influence (P. Punia, unpublished). Furthermore, the effect on ARMS specificity is also variable. The precise structures formed at the priming site are clearly important and it is not yet possible to formulate any rules that might guide the design process. However, it seems probable that mismatches close to the 3' terminal will cause the greatest effect on primer extension. Bases that are remote from the terminal exert their effects by destabilising the primer/template complex. This reduces the efficiency of priming by limiting the time that the complex is available to interact with the polymerase. It seems likely that primers with 3' ends rich in Gs and Cs are less likely to be destabilised by additional mismatches. Empirical evaluation of different structures in ARMS assays is required.

Design of the Common Primer

The primer paired with ARMS primers specific for the sequence variants is selected on the basis of the usual considerations used in primer design. The GC content should be approximately 50%, there should be no repeat or palindrome sequences and it should not be complementary with any of the specific primers.

Selection of PCR Conditions

Careful optimisation is required to maximise the specificity of ARMS. The tolerance of the PCR process to mismatches between the primer and the target sequence is dependent on the reaction conditions. Optimisation is aimed at maintaining the efficiency of matching template amplification while minimising the amplification of mismatching template.

PCR accommodates mismatched primers better when the annealing temperature is lower (Christopherson *et al.*, 1997). Variations in the length of time allowed for annealing also effect the efficiency of amplification by mismatched primers. The appropriate temperature and length of time used for the annealing step depends on the length, base composition and concentration of the amplification primers. Increasing annealing temperatures should have a greater destabilising effect on primers with lower melting temperatures. Raised annealing temperatures should therefore enhance the discrimination of ARMS reactions by reducing the extension of primers that do not match the template. Stringent conditions (*i.e.* annealing temperatures close to the primer:template duplex melting temperature) are essential for maximum ARMS specificity especially during the early cycles of PCR. The length of the annealing step hold time is of crucial importance since this is proportional to the likelihood of primer extension. Real-time machines such as the Lightcycler® that allow rapid cycling are extremely advantageous for ARMS PCR. These instruments allow the annealing temperature to be applied momentarily. This is followed by rapid transition to an extension temperature that is too high for significant primer annealing. This allows minimal time for the polymerase to extend any primers and in theory, mismatch extension is reduced to a negligible level.

Polymerases for ARMS

Thermostable DNA polymerases such as Vent and *Pfu* that have proof-reading exonuclease activity are unsuitable for ARMS PCR. These enzymes are able to remove 3' mismatched bases allowing extension to proceed from the matched n-1 base. Previous studies (Tada *et al.*, 1993) have reported that the Stoffel fragment of *Taq* DNA polymerase enhances discrimination in ARMS PCR. This enzyme lacks 5'-3' exonuclease activity and has lower processivity than native *Taq* polymerase. We have confirmed a modest increase (>1 $\log_{10}$) in ARMS specificity when using Stoffel fragment compared to the native enzyme (P. Punia, unpublished). The mechanism for this remains unclear but it is possible that it is related to the lower affinity of the modified enzyme for the primer/template complex which is also the reason for the impaired processivity of Stoffel.

The synthesis of extraneous products arising from non-selective primer/template hybridisation may occur during PCR set up at low temperatures, since thermostable polymerases retain partial activity under these conditions. *Taq* polymerase is active over a broad range of temperatures and primer extension occurs both during reaction set-up and during the temperature transition to the first denaturation hold. For efficient ARMS it is essential that these opportunities for mismatch extension are minimised. A 'hot-start' is therefore recommended. Classically 'hot-start' is used to reduce non-specific annealing of primers and minimise the amplification of non-specific products. The most efficient systems for ensuring that the polymerase remains inactive until the first denaturation step are to use either an anti-*Taq* antibody (TaqStart, Clontech, Palo Alto, CA) that abolishes polymerase activity or a modified polymerase such a TaqGold (Perkin Elmer, Boston) that requires heat activation. The activity of the antibody bound enzyme (*Taq* polymerase or Stoffel) is restored rapidly when the antibody is denatured by heat during thermal cycling. The modified *Taq* polymerases generally require exposure to heat for several minutes

218

for complete activation and this is achieved by extending the first denaturation hold, usually for several minutes.

Probes for Quantitative ARMS

In common with other quantitative real-time PCRs the reliability of quantitative ARMS is improved if a gene-specific probe is used to detect the amplicon. For the HBV ARMS described here the ResonSense® (Lee *et al.*, 2002) system is employed. The probe has a 5' Cy5 label and is biotinylated on the 3' end to prevent extension and competion with the primers for PCR reagents which could lead to false positive results. However, probe manufacturers have stated that biotinylation is only 95-99% efficient and that probe extension can occur. To overcome the problem, in addition to biotinylation, extra bases that mismatch the target can be added onto the 3' end of the probe. The 3' terminal bases act as a loose 'tail' ensuring that the 3' end of the probe does not bind to the target and eliminates any possibility of probe extension. The probe can vary in length and position within the target amplicon and can be complementary to either strand.

Potential Effects of the Low Fidelity of *Taq* Polymerase on ARMS Specificity

Specificity of ARMS depends on the extension or non-extension of the 3' selective primer. However, the effect of mis-incorporation of bases in the strand synthesised from the non-selective primer must be considered. The quoted infidelity rate of *Taq* polymerase in PCR is in the region of one base per ten thousand incorporated. The high rate of base misincorporation in PCR is determined by the rate of incorrect base addition, the rate of mismatched base extension and the time allowed for elongation. If this rate occurs during the first elongation step of the ARMS reaction and assuming an equal probability of all possible mismatches, it can be calculated that one template copy in every 4×10^4 replicated will have any particular base change. Thus after one elongation step starting with 4×10^4 identical copies of a wild-type template a single copy of any particular mutant will have been produced. On this basis, replication of $4x10^8$ copies of wild-type template would produce $1x10^4$ copies of any particular mutant template. This cannot be the whole story since as already discussed the efficiencies of extension of different mismatched bases varies very greatly. However, it is worth noting the theoretical limit that *Taq* fidelity places on the specificity of quantitative ARMS.

Detection of HBV Polymerase YMDD Motif Variants Associated with Resistance to Lamivudine

Lamivudine is an antiviral drug that is used to treat hepatitis B virus (HBV) infection. Treatment usually suppresses viral replication and interrupts liver damage. However, long-term therapy does not completely eliminate the virus and resistant mutants emerge in a large proportion of cases. Resistance is mediated by changes in the YMDD motif of the polymerase gene catalytic site (Allen *et al.*, 1998, Chayama *et al.*, 1998).

The YMDD (tyrosine-methionine-aspartate-aspartate) motif is situated centrally in the C domain of the DNA polymerase gene close to the putative lamivudine binding site and is highly conserved in drug naive virus. Two mutations are seen at this site. In the first, valine is substituted for methionine at residue rt204 (YVDD) and, in the second, the rt204 methionine is replaced by an isoleucine residue (YIDD). The YVDD mutation is usually accompanied by an upstream substitution of leucine (at residue rt180) by methionine. The rt180 mutation is also found with the YIDD variant but less frequently. There is good evidence that the rt204 mutation confers lamivudine resistance while the rt180 mutation, which usually appears later, is important in ameliorating the deleterious effect on virus replication rates. In summary, three sequence variants must be detected at residue 204, these are the wild-type (ATG) and the mutants (ATT and GTG). The variants at amino acid 180 are the wild type (TTG and CTG) and the ATG (mutant) (Pillay *et al.*, 1998) (Figure 2).

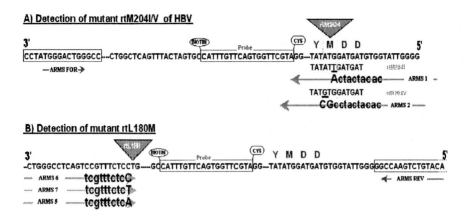

Figure 2. Diagram representing ARMS PCR method for the detection of mutants at residues a) rt204 and b) rt180. A) For the detection at rt204 a common forward primer on the sense strand is used in all reactions and the anti-sense primers have the terminating base corresponding to the mutant base on the HBV DNA. Primer to detect the GTG mutant also has an additional mismatch at the n-1 position to further destabilise PCR of wild-type DNA, without affecting amplification of mutant DNA. All reactions are done in parallel with the corresponding wild-type primers (not shown). B) Detection at rt180 is performed with the common primer designed to the anti-sense strand and three forward primers to detect the two known wild-types (CTG and TTG) and one mutant type (ATG). The CY5 labelled fluorescence probe is common for all reactions.

Real-time quantitative ARMS PCRs were developed for each of the key sequence variants. These assays are designed to measure the levels of each of the sequence types in the serum of patients treated with lamivudine. This information is useful in the management of antiviral therapy. For example, early warning of the emergence of lamivudine resistance might indicate the need for immediate changes to the drug regime including the introduction of combination anti-viral therapy.

The detection system for the HBV lamivudine resistance-associated mutations requires that data from two pairs of primers be employed in the quantification of each base (Figure 2). For example, the primer used to detect the ATT mutant has a terminal A base while the corresponding primer that detects the ATG wild-type has a C terminus. In the same way the primer for quantification of the GTG triplet ends in a C residue complementary to the G in the first position of the codon. This primer is paired with another that ends with a T base complementary to the first base of the ATG codon. A common primer, complementary to a conserved region of the virus, is used in all reactions.

HBV Mutation Detection Methods

HBV DNA Extraction from Serum

DNA for ARMS analysis can be extracted by any method known to be compatible with PCR. For HBV work serum samples (100µl) can be extracted using the Qiagen Viral DNA kit (Qiagen, Crawley, UK) according to the manufacturer's protocol. The DNA is eluted in 100 µl water.

ResonSense® Probe ARMS PCR Detection of YMDD Variants

The sequences of the probe and primers used for detection of the HBV mutants are given in Table 2. The primers amplify two separate loci (amino acids rt204 and rt180) associated with lamivudine resistance. The four ARMS primers that are directed at the rt204 codon each have different 3' ends. Two of these primers are complementary to the wild-type ATG codon with 3' ends at either the first or third base. The two mutant specific primers also target these positions. The three rt180 codon primers each target a different base at the first position of the codon. The rt204 and rt180 ARMS primer sets are complementary to different strands and consequently the 'common' primers are different at the two loci.

Reaction Conditions

Standard PCR buffers are suitable for use in quantitative ARMS assays. The HBV assays contained the following components: 50mM Tris-HCl pH8.3, bovine serum albumin (5µg/µl), 3mM MgCl$_2$, 400µM dNTPs, 2.5 pmol of each primer (Table 2), 2.5 pmol of Cy5-labelled probe, SYBR Green I at a final concentration of 1:10000 (BioGene

Table 2. Primers used for ARMS PCR for detection of mutations at residues rt204 and rt180.

Primers for detection of mutants at rt204		Primers for detection of mutants at rt180	
Primer common to all 204 reactions		Primer common to all 180 reactions	
ARMS FOR: CCTATGGGGAGTGGGCCTCAG		ARMS REV: GATGCTGTACAGACTTGGCC	
ARMS Primer	Codon detection	ARMS Primer	Codon detection
ARMS1: CCCCAATACCACACATCATCA	ATT	ARMS5: GCCTCAGTCCGTTTCTCA	ATG
ARMS2: CCCAATACCACACATCATCCGC	GTG	ARMS6: GCCTCAGTCCGTTTCTCC	CTG
ARMS3: CCCCAATACCACACATCATCC	ATG	ARMS7: GCCTCAGTCCGTTTCTCT	GTG
ARMS4: CCCAATACCACACATCATCCAT	ATG		

Cy5-labelled probe

5'CY5-TACGAACCACTGAACAAATGAATTG-Biotin3'

Ltd, Cambs, UK) and 5 units of Amplitaq DNA Polymerase Stoffel Fragment (Applied Biosystems, Foster City, CA) that had been pre-bound to 2.2µg Taqstart Antibody (Sigma-Aldrich Co, Dorset, UK) for 1hr at room temperature. HBV DNA or diluted first round PCR product (2µl) was added to each 10µl reaction.

Cycling Conditions

Denaturation at 94°C for 10 s was performed followed by; 40 cycles of 60°C for 0 s (fluorescence measurement), 74°C for 10 s and 92°C for 0 s. Melt analysis was done immediately following cycling by rapid cooling to 50°C and then increasing the temperature to 94°C at a ramp rate of 0.2°C per second with continuous fluorescence measurement.

Standards

A plasmid carrying a section of the polymerase gene including the rt204 and rt180 codons was purified by standard methods and linearised by PstI digestion. Linear plasmids were serially diluted 1:10 in 0.5 $\mu g/\mu l$ BSA (Bovine Serum Albumin, Life Technologies, Invitrogen Ltd, Paisley, UK) to give a dilution range of 5 to $5x10^9$ copies. To determine the selectivity level of the primers, reactions containing 10^9 copies of wild-type DNA template (ATG at codon 204) and from 10^9 to 0 copies of mutant virus template were tested. Two series of samples were prepared representing the ATT and GTG mutant in the wild-type background. When high levels of mutant sequence (10^9 copies – 10^5 copies) were mixed with a constant number of wild-type copies (10^9), the mutant sequence specific primers clearly detected its matching template as judged from a decrease in the crossing threshold. Mixed sample reactions containing 10^4 or fewer copies of the mutant template together with 10^9 copies of the wild-type template had indistinguishable crossing thresholds. Therefore, the mutant primers selectivity level was 1:10000 mutant:wild-type (Figure 3).

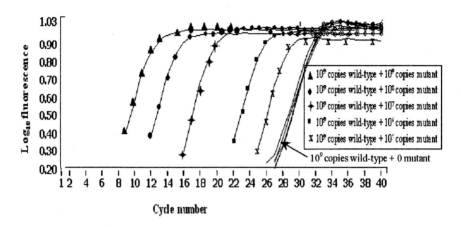

Figure 3. A mixture of the dilution series of mutants with fixed concentration of wild-type DNA was assayed with mutant primer to establish the concentration at which the mutant primers would detect the wild-type DNA. The above graph shows the log fluorescence versus crossing point cycle number.

Quantitative ARMS

For the quantitative ARMS assays the concentrations of the standards were chosen to match the expected concentration range of the samples. Standard curves were prepared using at least five standards. The DNA concentrations for the external standards were 5×10^9, 5×10^7, 5×10^6, 5×10^5, 5×10^3, 50 and 0 plasmid copies. Quantitative ARMS assays were performed in parallel reactions, one with a primer matching the variant at the 3' end, and the other with the primer matching the wild-type at the 3' end. All reactions were also run in parallel with standards of known DNA concentration.

Final Comments

In the HBV example, the rapid detection and quantitation of lamivudine resistant mutants can contribute to therapy management and the introduction of alternative treatments. The real-time ARMS PCR system is adaptable to detect other mutations, although the design of primers and simple PCR optimisation is required for each assay.

225

References

Allen, M.I., Deslauriers, M., Andrews, C.W., Tipples, G.A., Walters, K.A., Tyrell, D.L.J., Brown, N. and Condreay L.D. 1998. Identification and characterization of mutations in Hepatitis B virus resistant to lamivudine. Hepatology. 27: 1670-1677.

Ayyadevara, S., Tahnden, J.J. and Shmookler Reis, R.J. 2000. Discrimination of primer 3'-nucleotide mismatch by taq DNA polymerase during polymerase chain reaction. Anal. Biochem. 284: 11-18.

Billadeau, D., Blackstadt, M., Greipp, P., Kyle, R.A., Oken, M.M., Kay, N. and Van Ness, B. 1991. Analysis of B-lymphoid malignancies using allele-specific polymerase chain reaction: a technique for sequential quantitation of residual disease. Blood. 78: 3021-3029.

Chayama, K., Suzuki, Y., Kobayashi, M., Kobayashi, M., Tsubota, A., Miyano, Y., Koike, H., Kobayashi, M., Koisa, I., Arase, Y., Saitoh, S., Murashima, N., Ikeda, K. and Kumada, H. 1998. Emergence and takeover of YMDD motif mutant hepatitis B virus during long-term lamivudine therapy and re-takeover by wild-type after cessation of therapy. Hepatology. 27: 1711-1716.

Christopherson, C., Sninsky, J. and Kwok, S. 1997. The effects of internal primer-template mismatches on RT-PCR: HIV model studies. Nucleic Acids Res. 25: 654-658.

Day, P.J., Bergstrom, D., Hammer, R.P. and Barany, F. 1999. Nucleotide analogs facilitate base conversion with 3' mismatch primers. Nucleic Acids Res. 27: 1810-1818.

Huang, M. M., Arnheim, N. and Goodman, M. F. 1992. Extension of base mispairs by Taq DNA Polymerase; implications for single nucleotide discrimination in PCR. Nucleic Acids Res. 20: 4567-4573.

Kwok, S., Kellogg, D.E., Mckinney, N., Spasic, D., Goda, L., Levenson, C. and Sninsky, J.J. 1990. Effects of primer-template mismatches on the polymerase chain reaction: human immunodeficiency virus type I model studies. Nucleic Acids Res. 18: 999-1005.

Lee, M. A., Siddle, S. L. and Hunter, R. P. 2002. ResonSense: Simple Linear probes for quantitative homogeneous rapid polymerase chain reaction. Anal. Chim. Acta. 457: 61-70.

Newton, C.R., Graham, A., Heptinstall, L.E., Powell, S.J., Summers, C., Kalsheker, N., Smith, J.C. and Markham, A.F. 1989. Analysis of any point mutation in DNA. The amplification refractory mutation system (ARMS). Nucleic Acids Res. 17: 2503-16.

Okayama, H., Curiel, D.T., Brantly, M.L., Holmes, M.D and Crystal, R.G. 1989. Rapid, nonradioactive detection of mutations in the human genome by allele specific amplification. J. Lab. Clin. Med. 114: 105-113.

Pillay, D., Bartholomeusz, A., Cane, P.A., Mutimer, D., Schinazi, R.F. and Locarnini, S.A. 1998. Mutations in the hepatitis B virus DNA polymerase associated with antiviral resistance. Inter. Antiviral News. 6: 167-69.

Tada, M., Omata, M., Kawai, S., Saisho, H., Ohto, M., Saiki, R. and Sninsky, J.J. 1993. Detection of ras gene mutations in pancreatic juice and peripheral blood of patients with pancreatic adenocarcinoma. Cancer Res. 53: 2472-2472.

10

Real-Time NASBA

Sam Hibbitts and Julie D. Fox

Abstract

NASBA is an isothermal nucleic acid amplification method that is particularly suited to detection and quantification of genomic, ribosomal or messenger RNA. The product of NASBA is single-stranded RNA of opposite sense to the original target. The first developed NASBA methods relied on liquid or gel-based probe-hybridisation for post-amplification detection of products. More recently, real-time procedures incorporating amplification and detection in a single step have been reported and applied to a wide range of targets. Thus real-time NASBA has proved to be the basis of sensitive and specific assays for detection, quantification and analysis of RNA (and in one case DNA) targets. Molecular beacons have been utilised for detection of NASBA products in all published real-time procedures whether for commercially-available kits or for in-house diagnostic assays. As experience in design of such fluorescent-labelled probes increases and fluorimeters suitable for their detection become widely available, real-time NASBA methodology will be confirmed as a suitable alternative to other real-time amplification methods such as reverse transcriptase PCR (RT-PCR).

Introduction and Background to the Methodology

NASBA technology has provided an alternative method to standard procedures with a broad application for the amplification and detection of a range of nucleic acid targets (Compton, 1991). The majority of applications have been developed for detection and analysis of RNA targets including viral genomes, viroids, ribosomal RNA (rRNA) and messenger RNA (mRNA). Advantages of NASBA above methods such as RT-PCR include fast amplification kinetics and selective amplification of RNA in a background of DNA. The amplification is isothermal and the single-stranded RNA amplicons produced can be used directly in subsequent rounds of amplification or probed for detection without the need for denaturation or strand separation. Thus thermocycling is not required for NASBA.

Sample Preparation for NASBA

Amplification inhibitors and RNA integrity are the main cause of concern when preparing clinical specimens for NASBA. There is no need to degrade associated DNA if non-spliced mRNAs are to be analysed by a transcription-mediated amplification system such as NASBA since the standard method does not utilise temperatures which would denature and allow amplification of DNA. The method first detailed by Boom and colleagues (Boom *et al.*, 1990) is widely used for extraction of RNA for use in NASBA and reagents for this are commercially available. One advantage of this procedure is that dilute clinical samples can be concentrated during the extraction and the approach has been validated for a wide range of specimen types.

NASBA Primers and Probes

As for other amplification-based procedures, NASBA requires two target-specific oligonucleotides suitable for use as primers in the amplification phase. A region for probe-specific detection of amplified products also needs to be identified. Design of primers and probes for NASBA, whether for use in end-point or real-time detection methods, follows the same general rules and guidelines as those for other assays using nucleic acid amplification. The difference between NASBA and other amplification methods is that in the former method the T7 RNA polymerase promoter sequence needs to be included on one of the primers. Design criteria for NASBA primers and probes are discussed in detail below.

Amplification

NASBA methodology is transcription-based with isothermal, homogeneous amplification occurring in an analogous manner to the replication of retroviruses (Compton 1991; Chan and Fox, 1999). The amplification has a non-cyclic and cyclic phase and has proved to be useful in development of very sensitive diagnostic assays (Chan and Fox, 1999; Deiman *et al.*, 2002; Hibbitts and Fox, 2002). In the first stage the RNA template is reverse transcribed followed by a second-strand DNA synthesis and transcription of the resulting double-stranded DNA. The anti-sense RNA produced is then amplified in the cyclic phase of the reaction. For NASBA to occur avian reverse transcriptase (RT), ribonuclease H (RNase H) and T7 RNA polymerase together with two target specific primers are required. The process proceeds at 41°C typically over a 90 min time period with the major amplification product being anti-sense single-stranded RNA.

NASBA requires a single melt step at 65°C for the target directed primers to anneal. The three enzymes involved in the reaction are not thermostable and therefore are added after the melt step. The subsequent single temperature of 41°C required for the amplification process eliminates the need for thermocycling equipment used in standard molecular amplification techniques. In a modification of the standard procedure, NASBA can be utilised for detection of DNA targets in which thermocycling is required for the first one or two steps (depending if the target is single- or double-stranded).

Detection of NASBA Products

Methods for the detection of NASBA products have been reported using a probe-capture hybridisation and electrochemiluminescence (ECL) (Kievits *et al.*, 1991; van Gemen *et al.*, 1994; Chan and Fox, 1999). The principle of ECL is based on a voltage induced oxidation-reduction reaction involving an immobilised NASBA product. Detection occurs through amplicon hybridisation to a biotinylated target specific capture probe coupled to streptavidin coated magnetic beads and the whole complex is immobilised on an electrode by a magnet. In the NucliSens® Basic Kit format (bioMérieux Ltd) a generic ruthenium-labelled detection probe (supplied in the kit) is hybridised to the target amplicon. End point detection methods such as ECL have proved useful when setting up real-time NASBA in order to validate procedures. An example of comparative results for end-point and real-time NASBA are given in Figure 1.

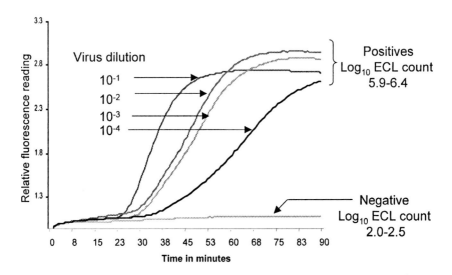

Figure 1. Example of results for real-time NASBA compared with an end-point detection method. Amplification of coronavirus dilutions was observed over a 90 minute time period. Real-time detection utilised a coronavirus-specific molecular beacon. ECL = electrochemiluminescence. Differentiation between positive and negative results is clear for both end-point and real-time detection methods.

Post-amplification detection procedures, such as ECL, require handling steps and machinery that affect both costs and turn-around time. The development of fluorescence-based assays combining amplification and detection for PCR and RT-PCR initiated studies to adapt NASBA methodologies into real-time procedures. Real-time NASBA enables product detection concurrent with target amplification in closed tubes, thus reducing both the handling steps and the risk of contamination in the post-amplification process.

All reported real-time NASBA procedures to date have utilised molecular beacons for detection of amplified products. Molecular beacons are hairpin shaped oligonucleotides with a loop region containing a probe sequence complementary to the amplicon and a stem with complementary arm sequences located on either end of the target sequence (Tyagi and Kramer, 1996). A fluorophore is covalently linked to one arm (5' end) and a quencher to the other (3' end). The quencher is a non-fluorescent chromophore that dissipates the energy it receives from the fluorophore as heat. However, when the probe sequence hybridises to its target it forms a rigid probe-target double helix that is longer and more stable than the stem hybrid. A conformational change occurs that separates the quencher from the fluorophore enabling fluorescence to be detected. Fluorescence increases concurrently with formation of amplified products by NASBA.

Leone *et al.* published the first study in which NASBA was combined with detection using molecular beacons (Leone *et al.*, 1998). Since then methodologies have been adapted to utilise real-time procedures for both kit-based and in-house assays (see Table 1).

Quantification of NASBA Products

Quantitative NASBA with end-point detection of amplified products is a well-established technique with internal calibrators of known copy number included in the reaction. Calibrators may be included at the beginning of the procedure ensuring they control all steps of the extraction, amplification and detection. A commercially available assay for quantification of HIV-1 by NASBA (with ECL detection)

Table 1. Example of design characteristics for real-time NASBA. Only published methodologies are included. Other assay details for unpublished/in preparation methodologies are available at www.basickit-support.com.

Reference and target(s)	Primer length (bases)		Molecular beacon (Dabcyl/Dabsyl used as quencher in each case)		
	P1 antisense*	P2 sense**	Target length (bases)	Stem length (bases)	Fluorescent Label(s)
Hibbitts et al. (2003) PIV1 and PIV3 (HN)	21 (both)	21 (both)	21 (PIV1), 20 (PIV3)	7 (both)	FAM
Greijer et al. (2002) CMV IE1 (WT and IC) and pp67	17 (pp67)	20 (pp67)	18 (pp67), 19 (IE1 WT and IC)	6 (all)	FAM, ROX, CB
Szemes et al. (2002) PVY (coat protein)	21	22	20–22	6 (all)	FAM, Texas red, HEX, TAMRA
Polstra et al. (2002) HHV8 (ORF 73, vGCR, vBcl-2 and vIL-6), U1A control (mRNA for snRNP)	24 (U1A) 25 (vBCl-2, vGCR) 27 (ORF 73) 28 (vIL-6)	23 (vIL-6, ORF 73, vGCR, U1A) 25 (vBcl-2)	22 (vBcl-2, vGCR and U1A), 23 (vIL-6), 24 (ORF 73)	6 (all)	FAM
van Beckhoven et al. (2002) Bacterial ring rot (16s rRNA)	18	18	20	6	FAM
De Baar et al. (2001a) HIV-1 (gag)	29	29	18-20	6 (all)	TET, ROX, FAM, TAMRA
De Baar et al. (2001b) HIV-1 (LTR)	18	20	20	6	FAM
De Ronde et al. (2001) HIV-1 (RT)	26	20	21	7 (both)	FAM, ROX

234

Table 1 continued.

Lanciotti and Kerst (2001) WN virus (envelope gene) and SLE virus (E glycoprotein)	23 (both)	22 (both)	22 (WN), 26 (SLE)	6 (WN), 7 (SLE)	FAM
Yates et al. (2001) Hepatitis B virus (single-stranded DNA region)	28	27	30	6	FAM

* P1 primer has the T7 RNA polymerase promoter sequence added to the target specific sequence at the 5' end.

** In some procedures P2 primer has a generic tail sequence used for end-point detection by ECL using the NucliSens® Basic Kit

$TCID_{50}$ = tissue culture infectious dose 50%, pfu =plaque forming units, RT = reverse transcriptase, LTR = long terminal repeat, HIV = human immunodeficiency virus, CMV = cytomegalovirus, IE = immediate early, WT = wild-type, IC = internal control, WN = West Nile, SLE = St Louis encephalitis, HN = haemagglutinin neuraminidase, HHV8 = human herpesvirus 8, ORF = open reading frame, ORF 73 = latency-associated nuclear antigen (LANA), vGCR = viral receptor, vBc1-2 = viral inhibitor of apoptosis, vIL-6 = viral growth factor, PVY = potato virus Y, PIV = parainfluenza virus, FAM = fluorescein, Dabcyl/Dabsyl = Dabcyl-200 or 4 (4'-dimethylaminophenylazo) benzoid acid, ROX= 6-carboxy-X-rhodamine, CB = Cascade blue, TET = tetrachloro-6-carboxyfluorescein, TAMRA = tetramethylrhodamine, ss = single-stranded, snRNP = small nuclear ribonucleoprotein, cfu = colony forming units.

uses three internal RNA calibrators (standards) which are added to the sample lysate prior to extraction (van Gemen *et al.*, 1994). These calibrators are amplified with the same primers as the wild-type target but have different internal sequences for probe detection. Assays set up in this way have a broad dynamic range (DR) and have proven clinical utility.

Developments in real-time NASBA led to modification of the HIV-1 assay. In the new HIV-1 NASBA, which utilises molecular beacons for product detection, a single calibrator was found to be sufficient to quantify target input (NucliSens EasyQ® HIV-1, BioMérieux Ltd). Weusten and colleagues described development of the mathematical model for this assay using two molecular beacons. The beacons comprised different loop structures and fluorophores that specifically bound to either HIV-1 sample RNA or calibrator amplicons. The fluorescence curves obtained for wild-type target and calibrator depend upon the NASBA-driven time-dependent growth in RNA levels and binding of the beacon to this RNA. As the concentration of the calibrator RNA in the reaction is known the levels of the HIV-1 sample RNA can be quantified using the developed mathematical model (Weusten *et al.*, 2002).

In recently-developed assays for in-house diagnostics time to positivity (TTP) has been utilised to estimate levels of nucleic acid target amplified in real-time NASBA. As can be seen in Figure 1 input copy number relates to development of detectable fluorescence in NASBA.

Multiplexing of Multiple Targets in a Single Reaction

Multiplex real-time NASBA assays have been reported with numerous molecular beacons present in one reaction. Different fluorophores can be measured simultaneously using defined excitation and emission filters. The commercially available HIV-1 assay utilises a simple multiplex reaction with molecular beacons specific for wild-type and calibrator amplicons having different labels (FAM and ROX, respectively). Other examples of multiplex reactions have been described using both end-point (*e.g.* van Deursen *et al.*, 1999; Hibbitts *et al.*, 2003) and real-time (de Baar *et al.*, 2001a; de Ronde *et al.*, 2001; Greijer *et al.*, 2002; Szemes *et al.*, 2002; Hibbitts *et al.*, 2003) detection. Thus the feasibility of mixing multiple primer and probe sets together has been shown.

Methods and Protocols for Real Time Assays

Design of Primers and Probes

A summary of design criteria used in successful real-time NASBA applications are given. Further details can be obtained from previous publications (Deiman *et al.*, 2002) and from a web site dedicated to NASBA applications (www.basickit-support.com). Optimisation of primer sets, the length and sequence of the molecular beacon and final concentrations of each in the reaction are critical to sensitivity and specificity of the resulting real-time NASBA assay.

The optimum length of amplification sequence has been quoted as approximately 100 to 250 nucleotides (including target-specific primer regions) although successful real-time NASBA assays have targeted up to 279 bases. As for other amplification assays, it is important that primer and probe sequences are checked for any cross-hybridisation or internal structure that could compromise the reaction. If possible, stretches of the same nucleotide should be avoided in primer design, the melt-temperature of the primers should be similar and the GC content of each close to 50%. The long length of some of the primers utilised in NASBA makes it particularly important that they are purified before use (HPLC or PAGE is usually used). The standard concentration of each primer used in NASBA (end-point or real-time detection) is 0.2 μM but a few investigators have seen benefit in adjusting the primer concentration, particularly in multiplex reactions.

The target-specific region of primer 1 (P1) is usually 20 – 25 nucleotides although sequences of 17 – 29 nucleotides have been utilised in real-time NASBA (Table 1). The 3' sequence of P1 complements the target nucleic acid and the 5' terminal has a T7 RNA polymerase promoter sequence tail (5' AATTCTAATACGACTCACTATAGGG). The transcription may be enhanced by an additional purine spacer region of 6-10 nucleotides added between the T7 polymerase promoter tail sequence and the target-specific region (*e.g.* AGAAGG) as described previously (Deiman *et al.*, 2002).

The second primer (P2) is complementary to the DNA sequence that is produced by extension from P1 at the 3' end. The target region for P2 is usually 20 – 25 nucleotides although variation outside this range is possible (Table 1). A generic sequence for ECL detection using the NucliSens® Basic Kit may be included at the 5' terminal (5'GATGC AAGGTCGCATATGAG). This tail sequence is not necessary if only real-time detection is to be used but does not seem to compromise detection using a molecular beacon.

The loop sequence of the molecular beacon designed to detect NASBA products should be homologous to the region of interest. The stem structure is provided by adding complementary residues to the 5' and 3' ends of the loop sequence. If the loop sequence is less than 25 nucleotides a 6 nucleotide stem (*e.g.* 5'CCAAGC...GCTTGG3') is usually used whereas loop sequences longer than 25 nucleotides may have a 7 nucleotide stem (*e.g.* 5'CCATGCG....CGCATGG3'). It is important that the probe designed should not form multiple structures. The stability and predicted structure of the beacon can be analysed by using the European MFOLD server (http://bibiserv.techfak.uni-bielefeld.de/mfold/). A name and sequence is input into the MFOLD enquiry form and an email address provided. Details of predicted RNA structures and thermodynamics are emailed and available up to 5 days post enquiry. In general, a free energy of -3 +/- 0.5 kcal/mole for the molecular beacon is recommended (Deiman *et al.*, 2002) although good structures with lower or higher energy values should still be considered. Reported real-time NASBA assays have utilised beacons with a loop varying between 18 and 30 bases. Deliberate mismatches and universal bases (*e.g.* inosine) can be used to alter probe binding and for discrimination between different RNA targets. A wide range of fluorescent labels has been utilised but in each case Dabcyl (or Dabsyl) has been used as a universal quencher. The concentration of molecular beacon in the real-time NASBA assay is generally between 0.5 μM and 0.05 μM.

Nucleic Acid Extraction Procedure

The following procedure describes use of the NucliSens® extraction kit (BioMérieux Ltd) which is one of the most widely evaluated procedures for preparation of RNA (and DNA) for NASBA and other amplification methods. It is based on the method first published by Boom and colleagues (Boom *et al.*, 1990).

As for all nucleic acid amplification procedures, specimen preparation should be carried out in a designated extraction area and gloves should be worn to prevent transfer of nucleases or contaminating nucleic acids (which can lead to false negative or false positive results, respectively). Sample is mixed with a guanidinium containing buffer in the proportion 1:9 (typically 100 µl of sample is added to 900 µl lysis buffer although this can be adjusted). Sample/lysis buffer mixtures are mixed and 50 µl of silica suspension added. After mixing, the tubes are held for 10 mins at room temperature. Care should be taken to vortex the tubes every few minutes to prevent the silica from settling to the bottom at this stage. Tubes are then spun in a microcentrifuge for 1 min to pellet the silica. The supernatant is carefully removed and wash buffer added. After mixing and pelleting, the wash procedure is repeated four times: once more with wash buffer; twice with 70% ethanol and once with acetone. The silica pellets are then dried in open test tubes for 10-15 mins at 56°C in a heating block. Purified nucleic acid (DNA and RNA) is eluted from the silica in an aqueous solution and residual silica removed by centrifugation. Extracted nucleic acid is then transferred to a fresh tube for storage (preferably – 80°C) before NASBA.

Real-Time NASBA using the NucliSens® Amplification Kit

NASBA reactions are carried out according to the general procedure reported previously (Kievits *et al.*, 1991) with some modifications for real-time analysis. The availability of reagents for NASBA in kit format makes the procedure easy to undertake but reagents and enzyme mixes can also be made in-house. The procedure described assumes that 10-11 tubes are to be utilised for analysis of target specific sequences in standard RNA NASBA. The Basic Kit manual can be used to modify the method if the kit-supplied performance control or other non-wild type control is also to be amplified.

The NucliSens® Basic Kit (BioMérieux Ltd) is used according to the manufacturer's instructions for standard amplification of RNA targets. Some minor changes are needed for amplification from DNA (*e.g.* Yates *et al.*, 2001). The kit contains the reagent sphere, diluent, stock KCl, nuclease-free water, lyophilised enzyme mix and enzyme diluent. For 10-11 amplification reactions, 55 µl of enzyme diluent is first added to the lyophilised enzyme mix and allowed to dissolve for 30 min at room temperature. 80 µl of the diluent is added to the reagent sphere and vortexed before addition of molecular grade water and stock KCl. Optimal KCl concentrations are determined for each target and are typically between 70-80mM. NASBA generally works well over a broad KCl concentration but this is one parameter that can be adjusted when working up a new procedure. The amplification solution is vortexed and 5 µl of each of the required primers (generally 5 µM stock of each although this can be varied) added along with the molecular beacon (generally 0.05 – 0.2 µM final concentration). The amplification solution is aliquoted into 10 µl portions in 1.5ml microtubes or strips of small tubes/wells depending on the fluorimeter to be used for real-time measurement of amplification. 5 µl of extracted nucleic acid is added to the amplification mix and extracted positive and negative controls are included in each assay. The reactions are incubated at 65°C for 2-5 min and then held for 2-5 min at 41°C. Subsequently, 5 µl of reconstituted NASBA enzyme solution is added to initiate amplification. The NASBA mix is usually incubated for a further 10 min in a 41°C heating block before transfer to a real-time fluorimeter. A commercially-available fluorimeter is available for detection of NASBA products (NucliSens® EasyQ analyser, BioMérieux Ltd) but any fluorimeter capable of holding samples at 41°C may be used for this procedure. Thermocycling is not required for real-time NASBA but where real-time PCR machines are available these may be utilised for NASBA (*e.g.* use of ABI 7700, Lanciotti and Kerst, 2001).

Table 2. Summary of performance characteristics for real-time NASBA. DR = dynamic range, TTP = time to positivity. Other abbreviations as for Table 1.

Reference (for target gene see Table 1)	Sensitivity/specificity	Multiplexing and typing	Quantification
Hibbitts et al. (2003)	Detection limit for each assay $\leq$ 1 TCID$_{50}$ or 100 RNA copies input. No cross-reaction with range of virus stocks.	Multiplex PIV1/PIV3 assay performed with no compromise in sensitivity and specificity.	Qualitative assays but linear relationship between input and TTP reported.
Greijer et al. (2002)	Detection limit 1–3 x 10^3 molecules / 100 μl blood. Specificity equivalent to monoplex (single target) assays.	3 label multiplex undertaken successfully.	DR of 10^3-10^6 RNA copies.
Szemes et al. (2002)	Detection limit 10-100 copies of RNA (100 copies in presence of all 4 beacons).	Tuber necrotic variants differentiated using 4 probes.	Qualitative assay.
Polstra et al. (2002)	Detection limit 50 copies input for each HHV8 target and 1000 copies input copies for U1A	Multiplex or typing assays not performed.	TTP used to calculate copy. Accurate quantification for HHV8 targets (100 – 1 x 10^7 copies input). Quantification of U1A over 10^3-10^8 copies input.
van Beckhoven et al. (2002)	Detection limit for optimised assay 10 –100 cfu input (100 cfu for complex starting material). Specificity checked on related bacteria.	Assay was used to identify viable cells.	Qualitative assay
de Baar et al. (2001a)	Detection limit 1000 copies input. 92% sensitivity on clinical panel.	HIV-1 type specific assays developed (A, B, C and recombinants)	Qualitative assay but linear relationship between input and TTP reported.
de Baar et al. (2001b)	Detection limit 10 copies input.	Assay not type specific.	TTP used to determine target amounts. DR 10^2-10^7 for M, N and O types.
de Ronde et al. (2001)	Not quoted.	Assay developed to identify RT variants.	DR good. Mixtures were identified efficiently with variants present at 1-4% detectable.
Lanciotti and Kerst (2001)	Detection limit 0.10 pfu input for WN virus and 0.15 pfu input for SLE virus. No cross-reaction with range of virus stocks.	Not applicable	Qualitative assay.
Yates et al. (2001)	Detection limit 10 input copies of DNA target.	Not applicable	Broad DR from 10^1 - 10^7

Applications of Real-Time NASBA

Leone *et al.* published the first study in which NASBA was combined with detection using molecular beacons (Leone *et al.*, 1998). The target for this assay was potato leafroll virus with detection sensitivity of 100-1000 synthetic RNA copies. Since then there have been many more reported applications, some of which have now been published in peer-reviewed journals (Table 1 and 2). Although many of the reported methods are set up as qualitative assays the time to positivity (TTP) can be used as an indication of RNA load in the sample. Yates *et al.* described the quantitative detection of hepatitis B virus DNA by real-time NASBA (Yates *et al.*, 2001). Thus although traditionally NASBA focuses on the amplification of RNA targets, in this study DNA was targeted with the production of RNA amplicons. The assay proved to be sensitive and had a wide DR. The authors reported that the reduced handling steps in real-time NASBA compared with alternative procedures resulted in a low risk of contamination. Other published methods have focussed on development of multiplex reactions or utilisation of real-time NASBA for typing and viability testing (Tables 1 and 2).

Where comparison has been made between different real-time procedures the rapid amplification of RNA by NASBA is apparent. In some cases, NASBA has proved to have superior sensitivity and or specificity above real-time RT-PCR (J.D. Fox and A. Rahman, unpublished; Lanciotti and Kerst, 2001) and thus should be considered as a suitable alternative amplification procedure.

Conclusions

Real-time NASBA assays are rapid, specific and sensitive with RNA amplification and target-specific detection achieved simultaneously and measured using a fluorimeter. Qualitative, quantitative, monoplex and multiplex formats of real-time NASBA have now been reported. The methodology has proved to be a suitable alternative to other amplification procedures without the need for expensive thermocyclers. An expansion of published methods using this alternative amplification and detection platform is likely over the next few years targeting genomic RNA, mRNA and rRNA.

References

Boom, R., Sol, C.J.A., Salimans, M.M.M., Jansen, C.L., Wertheim-van Dillen, P.M.E., and van der Noordaa, J. 1990. Rapid and simple method for purification of nucleic acids. J. Clin. Microbiol. 28: 495-503.

Chan, A.B., and Fox, J.D. 1999. NASBA and other transcription-based amplification methods for research and diagnostic microbiology. Rev. Med. Microbiol. 10: 185-196.

Compton, J., 1991. Nucleic acid sequence-based amplification. Nature 350: 91-92.

De Baar, M.P., van Dooren, M.W., de Rooij, E., Bakker, M., Van Gemen, B., Goudsmit, J., and De Ronde, A. 2001a. Single rapid real-time monitored isothermal RNA amplification assay for quantification of human immunodeficiency virus type 1 isolates from groups M, N, and O. J. Clin. Microbiol. 39: 1378-1384.

De Baar, M.P., Timmermans, E.C., Bakker, M., de Rooij, E., Van Gemen, B., and Goudsmit, J. 2001b. One-tube real-time isothermal amplification assay to identify and distinguish human immunodeficiency virus type 1 subtypes A, B, and C and circulating recombinant forms AE and AG. J. Clin. Microbiol. 39: 1895-1902.

De Ronde, A., van Dooren, M., van Der, H.L., Bouwhuis, D., de Rooij, E., Van Gemen, B., de Boer, R., and Goudsmit, J. 2001. Establishment of new transmissible and drug-sensitive human immunodeficiency virus type 1 wild types due to transmission of nucleoside analogue-resistant virus. J. Virol. 75: 595-602.

Deiman, B., Van Aarle, P., and Sillekens, P. 2002. Characteristics and applications of Nucleic Acid Sequence-Based Amplification (NASBA). Mol. Biotech. 20: 163-179.

Greijer, A.E., Adriaanse, H.M.A., Dekkers, C.A.J., and Middeldorp, J.M. 2002. Multiplex real-time NASBA for monitoring expression dynamics of human cytomegalovirus encoded IE1 and pp67 RNA. J. Clin. Virol. 24: 57-66.

Hibbitts, S., and Fox, J.D. 2002. The application of molecular techniques to diagnosis of viral respiratory tract infections. Rev. Med. Micro. 13: 177-185.

Hibbitts, S., Rahman, A., John, R., Westmoreland, D., and Fox, J.D. 2003. Development and evaluation of NucliSens® Basic Kit NASBA for diagnosis of parainfluenza virus infection with 'end-point' and 'real-time' detection. J. Virol. Methods. 108:145-55.

Kievits, T., van Gemen, B., van Strijp, D., Schukkink, R., Dircks, M., Adriaanse, H., Malek, L., Sooknanan, R., and Lens, P. 1991. NASBA isothermal enzymatic *in vitro* nucleic acid amplification optimised for the diagnosis of HIV-1 infection. J. Virol. Methods 35: 273-286.

Lanciotti, R.S., and Kerst, A.J. 2001. Nucleic acid sequence-based amplification assays for rapid detection of West Nile and St. Louis encephalitis viruses. J. Clin. Microbiol. 39: 4506-4513.

Leone, G., Van Schijndel, H., Van Gemen, B., Kramer, F.R., and Schoen, C.D. 1998. Molecular beacon probes combined with amplification by NASBA enable homogeneous, real-time detection of RNA. Nuc. Acids Res. 26: 2150-2155.

Polstra, A.M., Goudsmit, J., and Cornelissen, M. 2002. Development of real-time NASBA assays with molecular beacon detection to quantify mRNA coding for HHV-8 lytic and latent genes. BMC. Infect. Dis. 2: 18.

Szemes, M., Klerks, M.M., Van den Heuvel, J.F.J.M., and Schoen, C.D. 2002. Development of a multiplex AmpliDet RNA assay for simultaneous detection and typing of potato virus Y isolates. J. Virol. Methods. 100: 83-96.

Tyagi, S., and Kramer F.R.. 1996. Molecular beacons: probes that fluoresce upon hybridization. Nat. Biotechnol. 14: 303-308.

Van Beckhoven, J.R., Stead, D.E., and Van Der Wolf, J.M. 2002. Detection of *Clavibacter michiganensis* subsp. sepedonicus by AmpliDet RNA, a new technology based on real time monitoring of NASBA amplicons with a molecular beacon. J. Appl. Microbiol. 93: 840-849.

Van Deursen, P.B.H., Gunther, A.W., Spaargaren-van Riel, C.C., van den Eijnden, M.M., Vos., H.L., van Gemen, B., van Strijp, D.A.M.W., Tacken, N.M.M., and Bertina, R.M. 1999. A novel quantitative multiplex NASBA method: application to measuring tissue factor and CD14 mRNA levels in human monocytes. Nucleic Acids Research 27: e15.

Van Gemen, B., van Beuningen, R., Nabbe, A., van Strijp, D., Jurriaans, S., Lens, P., and Kievits, T. 1994. A one tube quantitative HIV-1 RNA NASBA nucleic acid amplification assay using electrochemiluminescent (ECL) labelled probes. J. Virol. Methods 49: 157-168.

Weusten, J.J., Carpay, W.M., Oosterlaken, T.A., van Zuijlen, M.C., and van de Wiel, P.A. 2002. Principles of quantitation of viral loads using nucleic acid sequence-based amplification in combination with homogeneous detection using molecular beacons. Nuc. Acids Res. 30: e26.

Yates, S., Penning, M., Goudsmit, J., Frantzen, I., Van de, W.B., Van Strijp, D., and Van Gemen, B. 2001.Quantitative detection of hepatitis B virus DNA by real-time nucleic acid sequence-based amplification with molecular beacon detection. J. Clin. Microbiol. 39: 3656-3665.

11

Applications of Real-Time PCR in Clinical Microbiology

Andrew David Sails

Abstract

The introduction of real-time PCR assays to the clinical microbiology laboratory has led to significant improvements in the diagnosis of infectious disease. There has been an explosion of interest in this technique since its introduction and several hundred reports have been published describing applications in clinical bacteriology, parasitology and virology. There are few areas of clinical microbiology which remain unaffected by this new method. It has been particularly useful to detect slow growing or difficult to grow infectious agents. However, its greatest impact is probably its use for the quantitation of target organisms in samples. The ability to monitor the PCR reaction in real-time allows accurate quantitation of target sequence over at least six orders of magnitude. The closed-tube format which removes the need for post-amplification manipulation of the PCR products also reduces the likelihood of amplicon carryover to subsequent reactions reducing the risk of false-positives. As more laboratories begin to utilise these methods standardisation of assay protocols for use in diagnostic clinical microbiology is needed, plus participation in external quality control schemes is required to ensure quality of testing.

Introduction

The first PCR methods to be described for clinical microbiology utilised gel electrophoresis for the detection of PCR amplification products. Although these assays proved useful, their specificity and sensitivity was compromised by this rather cumbersome end-point detection method. Specificity of detection could be improved by incorporating a solid phase hybridisation such as Southern blotting; however, this was labour intensive and time consuming requiring further manipulation of the PCR product. Detection of PCR products by solid phase hybridisation also limited the numbers of samples that could be processed, and the methods used were difficult to standardise between laboratories. The overall time taken to produce a result from a PCR assay could be two or three days and the test required a significant level of technical skill limiting the use of PCR to specialised laboratories only. The introduction of enzyme-linked hybridisation probe formats (PCR-ELISA) for the detection of amplification products did improve the detection process however they still required manipulation of the amplification products following PCR. Manipulation of the amplified product increases the likelihood of contaminating subsequent PCR reactions leading to false-positives a phenomenon known as amplicon carryover. PCR-ELISA facilitated the introduction of quantitative PCR (QPCR) assays however the range and accuracy of quantitation was limited. The more recent introduction of real-time platforms for PCR has revolutionised molecular diagnostic detection methods in clinical microbiology. These closed-tube systems virtually eliminate the risk of amplicon carryover because the samples are not opened following thermal cycling. Many of these new platforms process samples more rapidly than conventional block-based thermal cyclers making pathogen testing much more rapid. In addition, the ability to monitor the reaction in real-time provides results immediately after cycling and facilitates quantitation of the original target sequence over many orders of magnitude. Real-time platforms can differentiate between several closely related sequences within the same reaction therefore assays can be multiplexed to detect a range of pathogens within the same tube. Many of the assays described to date have utilised the Idaho LightCycler or the Roche LightCycler instrument, both of which I will refer to as LC for the purposes of this review. Some of the other commonly used platforms for real-time PCR are the Applied Biosystems ABI Prism 5700, 7000, 7700, and 7900 Sequence Detection Systems, and the Cepheid SmartCycler.

The real-time PCR method has been applied in virtually all areas of clinical microbiology and has proven useful in a wide range of applications. In this chapter I have provided an overview of some of the areas and applications where real-time PCR has made a significant impact in clinical microbiology and the diagnosis of infectious disease. This review is not intended to be exhaustive and does not attempt to describe every real-time PCR study which has been published therefore I would like to apologise in advance to authors of papers which I have overlooked in the preparation of this chapter.

Real-Time PCR Applications in Clinical Bacteriology

Real-time PCR assays have been described for a number of bacterial pathogens, some of which have been presented in Table 1. Some of the areas in which real-time PCR methods have made an impact on clinical bacteriology are described below.

Detection of Bacterial Respiratory Pathogens by Real-Time PCR

Bordetella pertussis causes whooping cough, a respiratory disease occurring mainly in adolescents. Laboratory diagnosis has traditionally relied on isolation of the organism by culture which is highly specific but sensitivity varies between 6 and 95% depending on the time of sampling (Kösters *et al*, 2002). The results of culture are also dependent on the quality of specimen provided. *B. pertussis* is a slow growing, fastidious organism and therefore isolation may take from three to

Table 1. Application of real-time PCR to the detection of clinically significant bacterial pathogens.

Organism	Gene targeted	Detection/ Quantitation	Sensitivity	Comments	Reference
Bacillus anthracis	*rpoB*	D	1 pg DNA/PCR		Qi *et al.*, 2001
B. anthracis	*rpoB*	D	25 spores/ml		Drago *et al.*, 2002
B. anthracis	*pagA, capB*	D	5 gene copies/PCR		Bell *et al.*, 2002
B. anthracis	*capB*	D	Not reported		Lee *et al.*, 1999
B. anthracis	*pagA, capB*	D	1 CFU/PCR		Makino *et al.*, 2001
B. anthracis	*rpoB, pagA, capC*	D	10 gene copies/PCR		Ellerbrook *et al.*, 2002
B. anthracis	*rpoB, lef*	D	50 GE^a/PCR	Multiplex	Oggioni *et al.*, 2002
B. anthracis	*cap, pag*	D	1 CFU/100 litres air	Multiplex	Makino *et al.*, 2003
Bartonella species	*ribC*	D	Not reported		Zeaiter *et al.*, 2003
Bordetella pertussis, B. holmesii	IS481	D	Not reported		Reischl *et al.*, 2001
B. pertussis, B. parapertussis	IS481, IS1001	D	0.1 CFU and 10 CFU/PCR		Kösters *et al.*, 2001
B. pertussis, B. parapertussis	IS481, IS1001, *ptg*	D	0.75 CFU/PCR	Multiplex	Sloan *et al.*, 2002
B. pertussis, B. parapertussis	IS481, IS1001	D	0.1 CFU and 10 CFU/PCR		Kösters *et al.*, 2002
B. pertussis, B. parapertussis	IS481, IS1001	D	1 CFU and 5 CFU/PCR	Multiplex	Cloud *et al.*, 2003
Borrelia burgdorferi	flagellin	Q	1-3 spirochaetes/PCR		Pahl *et al.*, 1999
B. burgdorferi	*recA*	Q	Not reported		Morrison *et al.*, 1999
Borrelia garnii, Borrelia afzelii, B. burgdorferi	*recA*	D	Not reported		Pietila *et al.*, 2000
Borrelia species	*ospA*	Q	1 to 10 spirochaetes/PCR		Rauter *et al.*, 2002

Table 1 continued

Organism	Target	Q/D	Sensitivity	Format	Reference
B. burgdorferi	flagellin	Q	1 to 10 spirochaetes/PCR		Piesman *et al.*, 2001
B. burgdorferi	flagellin	Q	1 to 10 spirochaetes/PCR		Zeidner *et al.*, 2001
B. burgdorferi	*recA*, p66	D	Not reported		Mommert *et al.*, 2001
Campylobacter species	16S rRNA	D	Not reported		Logan *et al.*, 2001
Campylobacter jejuni	Novel ORF[b]	Q	12 GE		Sails *et al.*, 2003
C. jejuni	Novel sequence[c]	Q	1 CFU/PCR		Nogva *et al.*, 2000
Chlamydia pneumoniae	*ompA*	D	0.001 IFU[d]/PCR		Tondella *et al.*, 2002
C. pneumoniae	*ompA*	Q	Not reported		Kuoppa *et al.*, 2002
C. pneumoniae	*ompA*	Q	10^{-6} IFU/PCR		Apfalter *et al.*, 2003
C. pneumoniae	16S rRNA	D	0.02 IFU/PCR		Reischl *et al.*, 2003b
C. pneumoniae	*Pst*I gene fragment	D	10 gene copies	Multiplex[e]	Welti *et al.*, 2003
Clostridium difficile	*tcdA*, *tcdB*	D	10 gene copies	Multiplex	Bélanger *et al.*, 2003
Escherichia coli (VTEC)	stx_1, stx_2	D	1 CFU/PCR	Multiplex	Bellin *et al.*, 2001
E. coli (VTEC)	stx_1, stx_2	D	10 gene copies	Multiplex	Bélanger *et al.*, 2002
E. coli (VTEC)	stx_1, stx_2, *eae*	D	5.8-580 CFU/beef, 1.2-1200CFU/faeces	Multiplex	Sharma *et al.*, 1999
E. coli (VTEC)	stx_1, stx_2, *eae*, *hlyA*	D	Not reported	Multiplex	Reischl *et al.*, 2002a
E. coli (VTEC)	stx_1, stx_2, *eae*[f]	D	Not reported	22 assays	Nielson and Anderson, 2003
E. coli (VTEC)	stx_1, stx_2, *eae*	D	10^4 CFU/g faeces		Sharma and Dean-Nystrom, 2003
Francisella tularensis	*fopA*, *tul4*	D	25 GE to 150 GE		Emanuel *et al.*, 2003

Table 1 continued

Organism		Gene target		Detection limit		Reference
Haemophilus influenzae	D	*bexA*		Not reported	Multiplex[g]	Corless *et al.*, 2001
Helicobacter pylori	Q	*ureC*		10 gene copies		He *et al.*, 2002
Legionella pnuemophila	D	*mip*		2.5 CFU/PCR		Ballard *et al.*, 2000
Legionella species, L. pnuemophila	D	5S rRNA, *mip*		<10 CFU/PCR		Hayden *et al.*, 2001
Legionella species, L. pnuemophila	D	16S rRNA, *mip*		1 fg DNA/PCR		Wellinghausen *et al.*, 2001
Legionella species, L. pnuemophila	D	16S rRNA		3 GE/PCR	Multiplex	Reischl *et al.*, 2002b
Legionella species	D	16S rRNA		2 CFU/PCR		Rantakokko-Jalava and Jalava, 2001
L. pnuemophila	D	*mip*		10 CFU/PCR		Wilson *et al.*, 2003
L. pnuemophila	D	*mip*		10 GE/PCR	Multiplex[e]	Welti *et al.*, 2003
Moraxella catarrhalis	D	*copB*		1 CFU/PCR		Greiner *et al.*, 2003
Mycobacterium tuberculosis	Q	IS6110		Not reported		Desjardin *et al.*, 1998
M. tuberculosis complex	D	ITS		800 gene copies/PCR		Kraus *et al.*, 2001
M. tuberculosis complex	D	ITS		Not reported		Miller *et al.*, 2002
Mycoplasma genitalium	Q	16S rRNA		10 gene copies/PCR		Deguchi *et al.*, 2002
M. genitalium	Q	16S rRNA		10 gene copies/PCR		Yoshida *et al.*, 2002
Mycoplasma pneumoniae	D	P1 gene		10 gene copies	Mulitplex[e]	Welti *et al.*, 2003
M. pneumoniae	D	P1 gene		Not reported		Hardegger *et al.*, 2000
Neisseria meningitidis	D	*ctrA*		Not reported	Mutiplex[g]	Corless *et al.*, 2001

Table 1 continued

				Genogrouping PCR	
N. meningitidis	16S rRNA, *sacC*, *siaD, porA*	D	Not reported		Molling *et al.*, 2002
Staphylococcus aureus	*nucA*	D	2 CFU/PCR		Palomares *et al.*, 2003
CNS[h]	16S rRNA	D	Not reported	Multiplex	Edwards *et al.*, 2001a
S. aureus, MRSA[i]	Novel sequence[j], *mecA*	D	25 GE/PCR	Multiplex	Reischl *et al.*, 2000
S. aureus, MRSA	Sa442[k], *mecA*	D	Not reported		Tan *et al.*, 2001
MRSA	Sa442, *mecA*	D	Not reported	Multiplex	Grisold *et al.*, 2002
MRSA	*nucA, mecA*	D	Not reported	Multiplex	Fang *et al.*, 2003
Group A Streptococci	not stated	D	Not reported		Uhl *et al.*, 2003
Streptococcus pneumoniae	*ply*	Q	1 CFU/PCR		Greiner *et al.*, 2001
S. pneumoniae	*lytA*	D	4 GE/PCR		McAvin *et al.*, 2001
S. pneumoniae	*ply*	D	Not reported	Multiplex[g]	Corless *et al.*, 2001
Tropheryma whipplei	16S 23S rRNA ITS, *rpoB*	Q	Not reported		Fenollar *et al.*, 2002
Yersinia enterocolitica	*bipA* and novel target sequence[l]	D	Not reported	Multiplex	Aarts *et al.*, 2001
Yersinia pestis	*pla*	D	2.1×10^5 gene copies/PCR		Higgins *et al.*, 1998
Y. pestis	*pla*, caf1, ymt, 16S rRNA	D	0.1 GE/PCR	Multiplex	Tomaso *et al.*, 2003

Table 1 continued

Notes:

a GE: genome equivalents

b 256 bp region of an open reading frame adjacent to and downstream from a novel two-component regulatory gene specific for *C. jejuni*

c 86 bp fragment including positions 381121 to 381206 of the published *C. jejuni* strain NCTC 11168 genome

d IFU: inclusion forming unit

e Multiplex assay including *C. pneumoniae*, *M. pneumoniae* and *L. pnuemophila*

f Multiplex PCR for *stx₁*, *stx₂*, *ehlyA*, *katP*, *espP*, *etpD*, *eae*, *tir*, *espD* (including all variants), and *saa*

g Multiplex assay for *N. meningitidis*, *H. influenzae*, and *S. pneumoniae*

h CNS: coagulase negative staphylococcus

i MRSA: methicillin resistant *S. aureus*

j Novel *S. aureus*-specific fragment

k Sa442: Novel *S. aureus*-specific fragment

l Novel 129 bp gene fragment specific for pathogenic strains of *Y. enterocolitica* twelve days. Reliable and rapid diagnosis facilitates the administration of appropriate treatment and prophylaxis of contacts. This has led to the development of a number of conventional PCR assays for the detection of *B. pertussis*. More recently a number of real-time PCR assays have been developed for the detection of *B. pertussis* in clinical samples. Reischl *et al.* (2001) described a LC assay for the detection of *B. pertussis* targeting the IS481 sequence. However, specificity studies demonstrated that the assay was also positive with strains of the closely related species *Bordetella holmseii* and sequencing of the PCR products revealed sequence homology between the two species. The authors concluded that the specificity and predictive value of IS481-based PCR assays might be compromised. A real-time 5' nuclease assay was described by Kösters *et al.* (2001) that targeted the IS481 sequence of *B. pertussis* and the IS1001 sequence of the closely related species *Bordetella parapertussis*. The assays demonstrated high sensitivity however similarly to the previous study, the species *B. holmseii* gave a positive signal in the IS481 assay. The assays were applied to 182 samples (nasopharyngeal swabs, nasopharyngeal aspirates [NPA], pharyngeal swabs, and tracheal swabs) from patients with and without symptoms of pertussis and the results compared to conventional culture. The real-time PCR assay demonstrated an increased sensitivity over culture with 28 patients being PCR positive/culture negative. Although 24 of these patients did meet the Centres for Disease Control and Prevention (CDC) clinical case definition for pertussis, the results must be interpreted with caution because *B. holmesii* also produced a positive signal in the assay in the specificity studies. Overall the PCR assay appeared to have a much greater sensitivity than culture with a detection rate which was nearly doubled compared to culture and which was similar to the previously reported studies using conventional PCR methods. A single-tube multiplex real-time LC assay was reported by Sloan *et al.* (2002) that also targeted IS481 and IS1001 in *B. pertussis* and *B. parapertussis*. Again the IS481 assay cross-reacted with four *B. holmseii* strains, similarly to previously reported assays. The authors considered that the sensitivity of using the IS481 as a target outweighed the limitations of specificity and reported that this assay has been adopted at their institution as the primary diagnostic test for this organism. The detection of *B. pertussis* and *B. parapertussis* by real-time PCR is a good example of where improvements in patient diagnosis have arisen from the introduction of PCR-based tests. In this case PCR-based diagnostic tests have replaced conventional methods because of their increased levels of sensitivity and the rapid timeframe in which results can be obtained. In the future real-time PCR assays for other slow growing pathogens and those which are difficult to recover from clinical samples by culture will be developed.

The slow growing respiratory pathogen *Mycobacterium tuberculosis* was also one of the first bacterial pathogens to be investigated by real-time PCR. Desjardin *et al.* (1998) developed a 5' nuclease QPCR assay targeting the IS6110 element specific for *M. tuberculosis*. They used this assay to determine if changes in the amount of *M. tuberculosis* DNA present in sputum correlated with the numbers of viable bacilli, therefore allowing the therapeutic response of patients to be monitored. They compared the results to acid-fast-bacilli (AFB) microscopy and culture. Prior to the start of therapy levels of AFB, *M. tuberculosis* DNA and cultivable bacilli were similar indicating that the assay may be useful for measuring the pathogen load prior to treatment. However, following the initiation of therapy levels of AFB and *M. tuberculosis* DNA did not correlate with cultivable bacilli indicating that the assay was not suitable for monitoring treatment efficacy.

Real-time PCR has also been applied to the direct detection of *M. tuberculosis* in clinical samples as a method to aid diagnosis. Kraus *et al.* (2001), developed a LC assay which targeted a 220 bp fragment of the ITS region of *M. tuberculosis*. Hybridisation probes were designed and the assay was demonstrated to be specific for members of the *M. tuberculosis* complex (*M. tuberculosis*, *Mycobacterium bovis* and *Mycobacterium africanum*). However the assay was not validated for the detection of these organisms directly in clinical samples. Miller *et al.* (2002) extended this work by applying the LC assay of Kraus *et al.* (2001) to the detection of *M. tuberculosis* in AFB smear-positive respiratory specimens and BacT/ALERT MP bottles. The LC assay demonstrated a sensitivity of 98.1% and 100% specificity for the AFB smear-positive samples and 100% sensitivity and specificity for 232 BacT/ALERT samples with 114 samples being positive in both culture and the LC assay. Real-time PCR assays may prove useful for the rapid confirmation of AFB smear-positive samples in the clinical laboratory however further studies are required to validate their sensitivity and specificity.

Legionella pneumophila is the causative agent of legionnaires disease; a nosocomial or community acquired pulmonary infection, which can be severe and life threatening. Infection occurs in immunocompromised patients and is acquired through inhalation of *L. pneumophila* from a contaminated environmental source, for example, the water system of large buildings such as hospitals. Rapid identification of the source of infection is important during outbreaks to limit further cases of infection. Occasionally other *Legionella* species such as *Legionella micdadei*, *Legionella bozemanii*, *Legionella dumoffii*, and *Legionella longbeachae* also cause opportunistic pneumonia. The isolation of *Legionella* from water samples is the current gold standard however it is limited by the organisms fastidious growth requirements, prolonged incubation periods (up to two weeks), the overgrowth of other bacteria and the presence of viable but non-culturable *Legionella*. Ballard *et al.* (2000) developed a LC biprobe assay specific for the macrophage infectivity potentiator (*mip*) gene of *L. pneumophila*. The limit of detection (LOD) of the assay was 2.5 CFU/reaction, equivalent to 1,000 CFU/litre of water. The assay was applied to the detection of *L. pneumophila* in 14 natural water samples and 10 laboratory microcosms of which 11 water samples and all 10 microcosms were culture positive for *L. pneumophila*. All 10 of the microcosms were positive in the LC assay however only six of the 11 culture positive water samples were positive in the assay. Three of the samples contained <200 CFU/litre and were below the detection limit of the assay and the other two samples contained PCR inhibitors. Overall the LC assay was a rapid method which has the potential for screening significant numbers of samples in a short time period. This may prove useful in outbreak situations where the reservoir of infection normally contains >1,000 CFU/litre.

A quantitative genus-specific LC assay was developed by Wellinghausen *et al.* (2001) for the detection and quantitation of *Legionellae* in hospital water samples. The assay targeted the 16S rRNA gene and included an internal inhibitor control based on a cloned fragment of lambda phage. A dual-colour hybridisation probe assay design was used and detected products were quantified with external quantitative standards (QS) composed of *L. pneumophila* DNA. The LOD of the assay was demonstrated to be approximately 1 fg of DNA (equivalent to one *Legionella* organism) and it detected all 44 *Legionella* species and serogroups. The assay was applied to 77 water samples from three hospitals with 54 of the samples being shown to be positive by culture. The LC assay detected *Legionella* in 76 of the 77 samples, with the quantitative results being generally 25-fold higher than those recovered by quantitative culture. These higher results may have been due to the assay detecting DNA from dead and non-culturable but viable cells. The authors developed a second LC assay based on the *mip* gene that was specific for *L. pneumophila* and re-tested the samples in the assay. All 76 samples were positive in the second assay correlating with the results of the first assay. The degree of contamination of hospital water supplies has been shown to correlate with the incidence of nosocomial infection, however the exact levels of contamination associated with infection have not been clearly established. The application of this quantitative assay to the monitoring of *Legionella* contamination levels in hospital water supplies may provide further data on the association between the level of contamination and risk of infection. This may facilitate the identification of critical levels of contamination useful for the monitoring of hospital water supplies thereby reducing the risk to patients.

The current gold standard for the diagnosis of *Legionella* infection is laboratory culture of the organism from clinical samples. Diagnosis can be made from a variety of clinical specimen types, however the bacterial culture of bronchoscopy or lung biopsy specimens is the most sensitive method of detection. Rapid diagnosis of infection in immunocompromised patients where the fatality rate can be as high as 50% is important because the infection often responds to antimicrobial therapy. The isolation of *Legionella* from patient clinical samples is hampered by the fastidious requirements of the organism and prolonged incubation periods of up to two weeks to ensure maximum recovery. Alternative more rapid tests, including direct fluorescent antibody assays (DFA) on respiratory secretions or urine antigen detection tests, have been developed however these too have some limitations. DFA can cross-react with other species leading to false-positive results and

DFA has a low sensitivity of detection. Antigen testing of urine has a relatively high sensitivity (85%) but can detect only a limited range of pathogenic *Legionella* species. Serological tests have been developed and are highly sensitive however their use is limited to epidemiological studies due to the significant time taken for patients to seroconvert. PCR-based diagnostic assays for the detection of *Legionella* species in clinical samples have been developed based on the detection of target regions in the 16S rRNA, 5S rRNA or the *mip* gene. Hayden *et al.* (2001) described a *Legionella* genus-specific LC assay targeting the 5S rRNA gene and a *L. pneumophila* specific LC assay targeting the *mip* gene for the detection of *Legionella* in clinical samples. The assays were applied to the detection of *Legionella* in bronchoalveolar lavage (BAL) and open lung biopsy samples from 35 patients with the samples being previously archived following conventional tests. The results of the LC assays were compared to culture, DFA and *in situ* hybridisation (ISH), the gold standard being culture. For the BAL specimens both the genus-specific and the *L. pneumophila*-specific LC assays demonstrated 100% sensitivity and specificity with nine of the samples being culture positive for *L. pneumophila*. The assays both demonstrated a reduced sensitivity for detection within the tissue samples, with the genus-specific PCR having a sensitivity of 68.8% and the *L. pneumophila*-specific assay having a sensitivity of 17%. This reduced sensitivity may have been due in part to the samples previously being formalin-fixed prior to recovery and testing. In addition, the *mip* gene is present as a single copy in the *Legionella* genome unlike the 5S rRNA which is present in multiple copies. These assays require further validation such as a prospective study to establish their utility for the detection of *Legionella* in clinical samples, especially tissue samples. Rantakokko-Jalava and Jalava (2001) described a LC SYBR Green I assay for the detection of *Legionella* species which targeted the 16S rRNA gene. This assay was applied to the detection of *Legionella* in 71 clinical samples from hospital patients with acute pneumonia, however only two samples were culture positive for *L. pneumophila* but they were positive in the PCR assay. No culture negative samples produced a positive result in the assay however the assay requires further application to clinical samples to validate its specificity and sensitivity for the detection of *L. pneumophila* in patient samples.

Reischl *et al.* (2002b) extended the previous study of Wellinghausen *et al.* (2001) by using the *Legionella*-specific primers and probes of the previous study in a LC assay for the detection of *Legionella* organisms in 26 culture-positive and 42 culture-negative BAL specimens. The study also utilised a second LC assay that targeted a *L. pneumophila* specific fragment of the 16S rRNA gene. Specificity studies revealed that some of the non-*L. pneumophila* species cross-reacted with the *L. pneumophila* probes and produced amplification plots, therefore melting curve analysis was required for unambiguous identification. The assay demonstrated 100% sensitivity and specificity for the BAL specimens tested, however it usefulness requires further investigation through prospective studies of samples from patients with atypical pneumonia.

The *L. pneumophila mip* gene was also targeted in the LC hybridisation assay of Wilson *et al.* (2003) which demonstrated a LOD of ten fg DNA equivalent to approximately ten organisms. The specificity of the assay was validated with a range of *L. pneumophila* strains, other *Legionella* species and species from unrelated genera. However, the authors did note that the assay might produce a positive result with *Legionella worsleinsis* and *Legionella fairfieldensis* due to the homology of the target sequence within these species and the target species *L. pneumophila*. The assay was applied to seven culture positive and 41 *L. pneumophila* culture negative clinical specimens with the assay results correlating with the culture results. Overall these real-time PCR assays may be a useful alternative to culture for the detection of *L. pneumophila* in clinical samples from patients with acute pneumonia, however further clinical trials are needed to validate this approach to the diagnosis of *Legionella* infection.

Chlamydia pneumoniae is an obligate intracellular pathogen that has been implicated as a cause of upper and lower respiratory tract infection in humans. It is also thought to be responsible for approximately 10% of cases of community-acquired pneumonia. Diagnosis is often based on serology or cell culture however these methods can produce inconclusive results. PCR detection methods may provide more rapid and reliable diagnosis of *C. pneumoniae* infection. Several real-time PCR assays have been described for the detection of *C. pneumoniae* many of which have targeted the major outer membrane protein gene (*ompA*) (Tondella *et al.*, 2002; Kuoppa *et al.*, 2002; Apfalter *et al.*, 2003), which is highly conserved within the species. Alternative gene targets have included the 16S rRNA gene (Reischl *et al.*, 2003b) and the *PstI* gene fragment of *C. pneumoniae* (Welti *et al.*, 2003). The assay of Kuoppa *et al.* (2002) was applied to the detection of *C. pneumoniae* in respiratory specimens with the results being compared to a conventional PCR assay. The real-time assay was demonstrated to be at least as sensitive for the detection of *C. pneumoniae* as an 'in-house' nested touchdown PCR, however the real-time assay was more rapid and less labour intensive. The assay of Reischl *et al.* (2003b) targeting the 16S rRNA gene was demonstrated to have a LOD of 0.02 inclusion-forming units (IFU) per PCR reaction equivalent to 1 IFU per ml of BAL. The assay was applied to 90 clinical samples from patients with pneumonia with 12 samples previously shown to be positive for *C. pneumoniae* also being positive in the real-time PCR assay. Although these results are promising further prospective studies are required to confirm the utility of this assay for the diagnosis of *C. pneumoniae* respiratory infections. Real-time PCR assays have been described for a number of other respiratory pathogens including *Mycoplasma pneumoniae* (Hardegger *et al.*, 2000), *Streptococcus pneumoniae* (Greiner *et al.*, 2001; McAvin *et al.*, 2001) and *Moraxella catarrhalis* (Greiner *et al.*, 2003).

Detection of Bacterial Meningitis by Real-Time PCR

Bacterial meningitis is a serious disease that affects the central nervous system (CNS) causing significant morbidity and mortality. Three pathogens are responsible for the majority of bacterial meningitis infections, these are *Neisseria meningitidis*, *Streptococcus pneumoniae*, and *Haemophilus influenzae*. The traditional method for the diagnosis of bacterial meningitis is the culture of the causative organism from cerebrospinal fluid (CSF) taken by lumbar puncture. However, culture can take up to three days to give a result and, following the practice of taking the clinical sample after the initiation of antimicrobial therapy the ability to recover the causative organism by culture has become more difficult. PCR-based diagnostic tests offer an alternative method to detect these organisms even after initiation of antimicrobial therapy, with bacterial DNA remaining detectable after the organism can no longer be recovered by culture. The first diagnostic PCR assays described were based on gel electrophoresis or PCR-ELISA end-point detection methods. Guiver *et al.* (2000) described the first real-time PCR assay for the detection of *N. meningitidis* in whole blood, CSF, plasma and serum samples. Assays were developed and evaluated targeting the meningococcal capsular transfer gene (*ctrA*), the insertion sequence IS1106 and two assays targeting the sialytransferase gene (*siaD*) for serogroup B and C determination. The specificity of the assays was investigated with a diverse range of meningococci serotypes and sero-subtypes plus species from other genera. The *ctrA* assay was demonstrated to be specific for *N. meningitidis* however the IS1106 assay produced false-positive results with non-meningococcal isolates. The *siaD* B and C assays exclusively detected serogroup B and serogroup C meningococci respectively with all other serogroups being negative in the assay. Application of the assays to tenfold dilutions of meningococci demonstrated that the LOD of the assays was less than one viable cell. The authors found that the adoption of the ABI 7700 platform increased the maximum number of samples that could be processed in a single day from 50 to 200 when compared to their current PCR-ELISA format. The authors adopted the *ctrA* assay as their primary screening test followed by serogroup determination using the *siaD* B and C assays. Following introduction of this assay as a routine test there was a 56% increase in the number of laboratory confirmed cases of meningococcal disease compared to culture only confirmed cases. This study is an excellent example of how the introduction of real-time PCR methods have significantly improved the recognition of the prevalence of meningococci in meningitis and is therefore an important model for other infectious diseases.

The *ctrA* assay was used in another study to determine if bacterial loads in meningococcal disease correlate with disease severity (Hackett *et al.*, 2002). The authors demonstrated that patients with meningococcal disease had higher bacterial loads than previously determined by quantitative culture. On hospital admission bacterial load was significantly higher in patients with severe disease with the maximum load being seen in patients who died. Corless *et al.* (2001)

developed a real-time multiplex assay for the simultaneous detection of *N. meningitidis*, *H. influenzae*, and *S. pneumoniae* in suspected cases of meningitis and septicaemia. The assay targeted the *ctrA* gene of *N. meningitidis*, the capsulation (*bexA*) gene of *H. influenzae* and the pneumolysin gene (*ply*) of *S. pneumoniae*. The meningococcal *ctrA* PCR demonstrated a sensitivity of 88.4% when tested against samples from culture-confirmed cases of meningococcal disease and the *H. influenzae bexA* assay demonstrated 100% sensitivity when tested against nine culture-confirmed cases of *H. influenzae* disease. The *S. pneumoniae ply* PCR assay demonstrated a sensitivity of 91.8% when applied to 36 samples from culture-confirmed *S. pneumoniae* disease. Co-amplification of two gene targets without a loss in sensitivity was demonstrated and the three primers and probe sets were combined into a multiplex assay. The multiplex PCR was then applied to 4,113 culture-negative clinical samples and cases of meningococcal, *H. influenzae* and pneumococcal disease that had not previously been confirmed by culture were identified. The application of this multiplex PCR in a routine screening service would provide improved non-culture diagnosis and case ascertainment of bacterial meningitis and septicaemia.

Detection of Borrelia Species by Real-Time PCR

Borrelia burgdorferi sensu stricto, *B. garinii* and *B. afzelii* are the genospecies of *Borrelia burgdorferi* sensu lato which cause Lyme disease (Lyme borreliosis), the most common tick-transmitted disease in humans in the Northern Hemisphere (Mommert *et al.*, 2001). *B. burgdorferi* sensu lato is the most common genospecies to cause infection in the United States with *B. garinii* and *B. afzelii* being the most prevalent genospecies in Europe. Infection results in a multi-system disease the most common symptoms being erythrema migrans (EM) and acrodermatitis chronica atrophicans (ACA). Neurological complications, arthritis and carditis can also occur. Diagnosis is usually based on clinical findings and serology although the spirochete can be cultured. However, these methods cannot quantitatively measure the spirochete load during infection, which would be useful for monitoring efficacy of therapeutic treatment and correlating the severity of symptoms with bacterial burden. Quantitative detection would also be useful to study bacterial burden in mouse models of Lyme borreliosis. A 5' nuclease assay targeting the flagellin gene was described for the quantitation of *B. burgdorferi* in mouse tissues following experimental infection and antibiotic treatment (Pahl *et al.*, 1999). The authors found a good correlation between clinical symptoms and spirochaete burden in the mouse model of Lyme disease. A second 5' nuclease quantitative assay was described which targeted the flagellin gene, which was applied to the analysis of spirochaete load in a mouse model of Lyme borreliosis. The authors concluded that the ability to monitor spirochete load during infection would allow the efficacy of therapeutic regimens to be investigated in the mouse model. Piesman *et al.* (2001) also used this assay to measure the density of *B. burgdorferi* in the midgut and salivary glands of feeding tick vectors. They identified that spirochaete load increased dramatically during feeding with a 6-fold increase in the tick mid-guts within 48h of attachment and feeding. Not surprisingly this time period coincides with the maximum increase in transmission risk during tick feeding.

Morrison *et al.* (1999) described a LC assay based on the *recA* gene that they also applied to the quantitation of *B. burgdorferi* in infected mouse tissues. The assay was demonstrated to be quantitative between 10 and 10^6 bacteria and was useful for measuring bacterial burden in various mouse tissues. This assay was utilised in two further studies to quantify levels of bacteria in human skin biopsies from patients with suspected Lyme borreliosis (Liveris *et al.*, 2002) and in field-collected *Ixodes scapularis* ticks in North-eastern United States which transmit infection to humans (Wang *et al.*, 2003). In the human skin biopsy study 40 of 50 skin samples from patients with EM were positive in the PCR assay with the number of spirochaetes detected varying between 10 and 11,000 per 2-mm biopsy with no association between the patient symptoms and numbers of spirochaetes. The primers described by Morrison *et al.* (1999) were also utilised in a LC assay for the differentiation of *B. garinii* from *B. afzelii* and *B. burgdorferi* based on melting curve analysis following amplification. Unfortunately differentiation between the latter two genospecies was not possible because they had indistinguishable melting curves leading Mommert *et al.* (2001) to look to alternative gene targets to provide differentiation.

They targeted the p66 gene in a nested LC PCR which also appeared to be more sensitive than the *recA* assay and provided differentiation between the *B. garinii* and *B. afzelii* genospecies, however the *recA* assay was required to distinguish between all three genospecies. These real-time assays may also be useful for the qualitative detection of *Borrelia* species in human clinical samples.

Detection of Bacterial Gastrointestinal Pathogens by Real-Time PCR

A large number of serotypes of verotoxin-producing *Escherichia coli* (VTEC) have been identified as the causative agents of haemorrhagic colitis and haemolytic ureamic syndrome. The *E. coli* O157:H7 serotype has been responsible for both sporadic cases and large outbreaks of infection and it can be isolated relatively easily using a selective-differential agar medium, sorbitol MacConkey agar. However, increasing numbers of strains with other serotypes are also becoming associated with serious human illness, including serotypes O26, O111 and O145. Unfortunately there is no selective medium that is suitable for the isolation and recognition of strains with these serotypes. Alternative detection methods have been developed to detect these additional serotypes based on conventional PCR with various gene targets including the Shiga toxin genes (stx_1 and stx_2). The first reported real-time assay was described by Oberst *et al.* (1998) who developed a 5' nuclease assay for the presumptive detection of *E. coli* O157:H7. Sharma *et al.* (1999) also developed a multiplex 5' nuclease assay for the detection of *E. coli* O157:H7 and additional serotypes of toxigenic *E. coli* targeting the stx_1, stx_2 and *eae* genes in a single reaction. The sensitivity of the assays in spiked beef and faeces were 5.8 to 580 CFU and 1.2 to 1200 CFU, respectively. The first LC assay to be described also targeted the stx_1 and stx_2 genes and utilised a hybridisation probe format with a duplex PCR reaction (Bellin *et al.*, 2001). They applied the assay to 48 VTEC and 37 *stx* negative strains demonstrating both its specificity and its ability to distinguish between strains that were positive for stx_1, stx_2 or both genes. The primers were designed based on a sequence alignment of 27 published stx_1 and stx_2 sequences with the stx_1 primers having a 100% match to all 27 sequences. The stx_2 primers contained two mismatches within all of the stx_{2e} sequences that facilitated differentiation between the $stx_{2/2v}$ and stx_{2e} alleles by melting curve analysis. A pair of LC duplex real-time PCR assays were described by Reischl *et al.* (2002a) that targeted the stx_1 and stx_2 genes in one reaction and the intimin (*eae*) and enterohemolysin gene (E-*hly*) in the second. Previous studies have demonstrated that strains of VTEC associated with human disease are positive for the *eae* and E-*hly*. The specificity of the assays was determined by applying them to 431 VTEC isolates, 73 *E. coli* strains negative for *stx* genes and 118 isolates of other species. The sensitivity and specificity of the LC assays were 100% for the stx_1, *eae*, and E-*hly* genes and 96 and 100% respectively for the stx_2 gene. No stx_2 genes were detected in ten stx_{2f} positive strains because of significant numbers of mismatches between the target gene sequences and the primers. The analytical sensitivity of the assays was relatively low at 5×10^2 CFU/PCR for the *eae* and E-*hly* genes and 5×10^2 to 5×10^3 CFU/PCR for the stx_1 and stx_2 genes. Application of the assays to spiked human stool samples demonstrated a similar level of sensitivity (5×10^3 CFU/ml stool suspension); however, the samples had to be cultured overnight on MacConkey agar. The apparent lack of sensitivity of the assays may limit their use for the detection of VTEC directly in stool samples from human patients.

Bélanger *et al.* (2002) developed a SmartCycler assay for the detection of stx_1 and stx_2 genes in VTEC using a multiplex reaction with a molecular beacon detection format. It demonstrated an analytical sensitivity of 10 genome copies per reaction, however when it was applied to the detection of VTEC in spiked human faeces the sensitivity was reduced at about 10^5 CFU/g faeces. Application of the PCR assay to 38 human faecal samples from VTEC infected and non-VTEC infected individuals demonstrated that the results correlated very closely to culture with a sensitivity of 100% and a specificity of 92%. One faecal sample was positive in the PCR but negative by culture indicating that the assay may be more sensitive than culture, however further prospective studies are required to confirm this finding. A real-time QPCR assay for VTEC was described by Sharma and Dean-Nystrom (2003) which targeted the stx_1, stx_2, and *eae* genes. The detection of the three genes was shown to be linear over a range of

10^4 to 10^8 CFU/g faeces for *E. coli* O157:H7 with a LOD in cattle and pig faeces of 10^4CFU/g. This assay requires further validation using naturally contaminated samples however it does appear promising as a tool for the rapid screening of faecal samples for VTEC. In addition it may also prove useful for the direct quantitation of VTEC in foods or faecal samples.

Clostridium difficile is the main etiological agent of antibiotic diarrhoea, pseudomembranous colitis and nosocomial diarrhoea. Infection with *C. difficile* is usually associated with healthcare settings where up to 20% of patients are asymptomatic carriers. Patients become predisposed to colonisation with *C. difficile* following disturbance of their normal intestinal flora by the use of antibiotics that can then lead to nosocomial diarrhoea. The pathogenesis of *C. difficile* disease is associated with two toxins, toxin A (TcdA) and toxin B (TcdB) which may cause mucosal damage. The gold standard for the diagnosis of *C. difficile* disease is the demonstration of the toxins in faecal samples using the tissue culture cytotoxicity assay. However, this assay requires between 24 and 48 hours to produce a result. Rapid enzyme immunoassays are available however they lack sensitivity. A number of conventional PCR assays have been developed for the detection of *C. difficile* however they require post-amplification manipulation of the product. Bélanger *et al.* (2003) described a SmartCycler-based real-time assay for the detection of both toxin genes in a multiplex molecular beacon format. The assay demonstrated a detection limit of approximately 10 genome equivalents per PCR using purified DNA and a limit of approximately 5×10^4 CFU/g faeces. The assay was applied to the detection of *C. difficile* in 56 faecal samples from hospital patients with the results being compared to the cytotoxicity assay. Twenty-nine faecal samples were positive in the cytotoxicity assay and 28 were positive in the real-time PCR assay with 27 samples being negative in both. Overall these preliminary results are promising and the assay may provide more rapid results than current diagnostic tests, however its utility needs to be validated in a larger clinical trial.

Real-Time PCR Applications in Clinical Parasitology

Real-time PCR methods have been utilised in several areas of clinical parasitology (Table 2). Examples include real-time PCR methods for the detection of faecal parasites which are more sensitive and specific than the conventional methods of microscopy or serology. Real-time PCR assays also facilitate sensitive and accurate quantitation of parasitic burden during infection even at very low levels and therefore may be useful for monitoring the efficacy of vaccines at pre-symptomatic levels. This ability to quantify very low-levels of parasitaemia may also be useful to assess the efficacy of new treatments during drug-trials and for diagnosing low level infections or carrier states. In the future real-time PCR assays will be developed for other endemic and emerging parasitic diseases leading to improvements in the prevention, diagnosis and treatment of these important human pathogens. Some of the areas of clinical parasitology in which real-time PCR assays have been introduced are reviewed below.

Detection of Enteric Parasites by Real-Time PCR

Cryptosporidia are coccidian protozoan parasites that infect a range of vertebrate hosts. More than 10 species of *Cryptosporidium* have been described and within the species group *Cryptosporidium parvum* there are also different genotypes. A number of species and *C. parvum* genotypes have been demonstrated to cause disease in humans, *C. parvum* human genotype being the most prevalent, with *C. parvum* bovine genotype, *Cryptosporidium meleagridis*, *Cryptosporidium felis* and *Cryptosporidium canis* (in order of prevalence) also causing disease. These organisms cause a self-limiting diarrhoeal disease in immunocompetent individuals; however, infection in immunocompromised patients can cause a severe life threatening disease with prolonged diarrhoeal illness. Conventional diagnosis is based on identification of oocysts in stained faecal smears however this lacks sensitivity and specificity and subsequent species or genotype identification following detection is not possible. A real-time 5' nuclease QPCR assay was described by Higgins *et al.* (2001), for the Cp11 and

Table 2. Application of real-time PCR to the detection of clinically significant parasites.

Organism	Gene targeted	Detection/ Quantitation	Sensitivity	Reference
Cryptosporidium species	16S rRNA	D	5 oocysts/PCR	Limor et al., 2002
Crytposporidium parvum	Cp11, 18S rRNA	D	not reported	Higgins et al., 2001
C. parvum	β-tubulin, GP900/poly(T)	D	1 oocyst/PCR	Tanriverdi et al., 2002
Cyclospora cayetanensis	18S rRNA	Q	1 oocyst/PCR	Varma et al., 2003
Entamoeba histolytica, Entamoeba dispar	rRNA	D	0.1 parasite/g faeces	Blessmann et al., 2002
Plasmodium falciparum	18S rRNA	Q	20 parasites/ml blood	Hermsen et al., 2001
P. falciparum, P. vivax, P. malariae, P. ovale	16S rRNA	D	0.1 parasite/PCR	Lee et al., 2002
Plasmodium yoelii	18S rRNA	Q	0.5 parasites/PCR	Witney et al., 2001
Encephalitozoon intestinalis	16S rRNA	Q	10^2-10^4 spores/ml faeces	Wolk et al., 2002
Enterocytozoon bienusi	16S rRNA	Q	10 GE[a]/μl stool	Menotti et al., 2003
Leishmania donovani and L. brasiliensis complexes	18S rRNA	Q	94.1 parasites/ml blood	Schulz et al., 2003
Leishmania infantum	DNA polymerase	Q	5 GE/PCR	Bretagne et al., 2001
Leishmania species	kDNA (kinetoplast)	Q	0.1 parasites/PCR	Nicolas et al., 2002
Toxoplasma gondii	B1 gene	Q	0.05 tachyzoite/PCR	Lin et al., 2000
T. gondii	B1 gene	Q	0.75 parasites/PCR	Costa et al., 2000
T. gondii	ITS1 of 18S rRNA	Q	1 bradyzoite	Jauregui et al., 2001
T. gondii	B1 gene and a multicopy genomic fragment	Q	2.5 GE/PCR	Reischl et al., 2003a
Trypanosoma cruzii	195 bp repeat[b]	Q	0.01 parasites/PCR	Cummings and Tarleton, 2003.

Notes:
[a] GE: genome equivalents
[b] 195 bp repeat unique to T. cruzii

18S rRNA genes of *C. parvum*. The assay was applied to the detection of oocysts in both human, cow and calf faecal samples however the probe hybridised to both human-infective and non-infective species limiting its usefulness. A LC hybridisation probe assay was described by Limor *et al.* (2002) for the detection and differentiation of most *Cryptosporidium* species and *C. parvum* genotypes. The assay had a high sensitivity (five oocyst detection level) and differentiated between species and genotypes by melting curve analysis. Although this assay appears useful it requires further validation through the application of it to human faecal samples from patients with *Cryptosporidium* infection. A 157-bp fragment of the β-tubulin gene of *C. parvum* was targeted by the real-time LC assay of Tanriverdi *et al.* (2002). The assay had a LOD of a single purified oocyst per reaction and was the first assay that could differentiate between *C. parvum* genotypes 1 and 2 on the basis of melting curve analysis.

Cyclospora cayetanensis is a coccidian parasite that has recently become recognised as an emerging pathogen of humans. *C. cayetanesis* causes prolonged diarrhoea, nausea, abdominal cramps, anorexia and weight loss. Transmission of infection occurs through environmentally resistant oocysts that are shed in the faeces of infected individuals. Similarly to other enteric parasites, diagnosis is dependent on faecal microscopy which lacks sensitivity and specificity. A 5' nuclease assay was described which targeted the 18S rRNA sequence of *C. cayetanesis,* and was demonstrated to be sensitive and specific for the detection of *C. cayetanesis* oocysts (Varma *et al.*, 2003). The LOD of the assay was 1 oocyst per 5μl reaction however the application of the assay to faecal samples from patients has not been reported.

Entamoeba histolytica is an intestinal protozoan parasite which is endemic in many parts of the world and is responsible for millions of cases of dysentery and liver disease per year (Blessmann *et al.*, 2002). Laboratory diagnosis of *E. histolytica* infection has been based on faecal microscopy of fresh or fixed stool samples. This method is hampered by the existence of the closely related but non-pathogenic species *Entamoeba dispar* that is morphologically indistinguishable from *E. histolytica*. Conventional PCR methods have been described for the detection and differentiation of these two species however these are time consuming and have limited throughput. Blessmann *et al.* (2002) developed a LC assay which targeted a 310-bp fragment of the high copy number ribosomal DNA. The assay utilised two sets of primers, each specific for one of the two species although the rDNA of the two species has a 98.4% sequence homology. The rDNA resides on an episomal plasmid that has a high copy number per cell (approximately 200 copies) which gave the assay a high sensitivity and a LOD of 0.1 parasite/g faeces. This represents a significant increase in sensitivity of detection when compared to microscopy, which has a sensitivity of approximately 70% if a single faecal smear is examined.

Microsporidia are obligate intracellular parasites that cause chronic diarrhoea in human immunodeficiency virus (HIV) infected patients and in other immunocompromised populations. *Enterocytozoon bieneusi* is the most commonly encountered species in humans (Menotti *et al.*, 2003); however other species have also been associated with infection including *Encephalitozoon intestinalis* (Wolk *et al.*, 2002). Diagnosis of intestinal infection is usually performed by faecal microscopy with stains such as the modified trichrome stain or Uvitex 2B, however such methods lack sensitivity and cannot distinguish between species. A number of conventional PCR assays have been developed for the detection of these pathogens but they are not quantitative and are laborious. A LC PCR assay based on hybridisation probes for the detection and quantitation of *Encephalitozoon* species in faecal samples was described by Wolk *et al.* (2002). Melting temperature analysis differentiated the three *Encephalitozoon* species (*E. intestinalis*, *E. cuniculi*, and *E. hellem*). The assay targeted a 268 bp region of the 16S rRNA gene and demonstrated a sensitivity of 10^2 to 10^4 spores/ml faeces compared to the sensitivity of trichrome blue stain of $\geq 1.0 \times 10^6$ spores/ml faeces. A real-time QPCR assay for the detection of *E. bieneusi* DNA in stool specimens from immunocompromised patients with intestinal microsporidiosis was described by Menotti *et al.* (2003). Similarly to the assay of Wolk *et al.* (2002) this 5' nuclease assay targeted the 16S rRNA gene of *E. bieneusi* and provided a quantitative range of detection of 10^1 to 10^7 copies of *E. bieneusi* DNA with a LOD of 10 copies. This assay was applied to the detection and quantitation of *E. bieneusi* DNA in the stools of patients in a randomised comparative trial of the efficacy of fumagillin for the treatment of microsporidial intestinal disease in the immunocompromised. The parasitic burden (*E. bieneusi* DNA in stool) remained stable in the placebo group (n=6) however the parasitic burden in patients treated with fumagillin (n=6) dropped to below the limit of detection. The real-time PCR assay performed better than semi-quantitative assessments by microscopy to measure parasitic burden.

Detection and Quantitation of Malaria Parasites by Real-Time PCR

Malaria is a major global health problem causing up to 500 million clinical cases and 2.7 million deaths annually. Malaria is caused by a parasitic protozoan of the genus *Plasmodium*. The gold standard for malaria diagnosis is the microscopic examination of a blood smear, however this may lack sensitivity when patients have low level parasitaemias. In addition, blood smear analysis cannot provide a quantitative assessment of the level of parasitaemia in the patient which may be useful for monitoring the effectiveness of anti-malarial therapy by measuring reductions in parasitaemia. Hermsen *et al.* (2001) described a real-time 5' nuclease QPCR assay targeting the 18S rRNA gene of *Plasmodium falciparum*. The assay was used to measure parasitaemia in five experimentally infected volunteers and the results compared to microscopic analysis of blood smears. The numbers of parasites detected by blood smear showed a good correlation with the real-time PCR assay. Similarly, Witney *et al.* (2001) developed a 5' nuclease QPCR assay for detection of the 18S rRNA gene of *Plasmodium yoelii*. They used the assay to detect and quantify liver stage parasites in experimentally infected mice as a means of evaluating the efficacy of pre-erythrocytic malaria vaccines. The authors demonstrated the assay to be a rapid and reproducible way of accurately measuring liver stage parasitic burden and vaccine efficacy in rodent malarial models. Real-time QPCR assays for malarial parasites may also be useful for monitoring the effectiveness of anti-malarial therapy during anti-malarial drug trials (Lee *et al.*, 2002).

Detection of *Toxoplasma gondii* by Real-Time PCR

Toxoplasma gondii is a protozoan parasite that has recently emerged as one of the most important opportunistic pathogens affecting AIDS patients and other immunosuppressed individuals. Infection can be transmitted to humans by ingestion of *T. gondii* oocysts in food, water or by consumption of tissue cysts in raw or undercooked meat. Pork is thought to be a significant source of human infection in the USA although infected lamb also contributes to infection world-wide and between 15 and 85% of the human population are thought to be infected asymptomatically. *T. gondii* causes toxoplasmic encephalitis and extracerebral toxoplasmosis, which are serious life threatening diseases in these patients. Toxoplasmosis infection during pregnancy may lead to serious or fatal disease of the foetus. Rapid diagnosis of infection in these two patient groups is required to allow timely initiation of treatment. Diagnosis of toxoplasmosis is based upon serological detection of specific anti-toxoplasma immunoglobulin, or culture of amniotic fluid or foetal blood. However, serological tests are not always reliable indicators of active infection because reactivation is not always accompanied by changes in antibodies. This has led to the development of QPCR assays for the sensitive, rapid and specific detection of *T. gondii* in clinical samples. Reported assays have targeted various genes including the *T. gondii* B1 gene (Lin *et al.*, 2000; Costa *et al.*, 2000; Reischl *et al.*, 2003a), a 529 bp repetitive fragment unique to *T. gondii* (Reischl *et al.*, 2003a), the 18S rRNA gene (Schulz *et al.*, 2003; Kupferschmidt *et al.*, 2001), and the ITS region of the 18S rRNA gene (Jauregui *et al.*, 2001). The 5' nuclease assay of Lin *et al.* (2000) targeted the B1 gene and had a LOD of 0.05 tachyzoites and a linear range of quantitation of 6 logs. The authors applied the assay to the detection of *T. gondii* in 30 paraffin-embedded foetal tissue sections. Ten of the samples were positive in the assay, which was consistent with the results of the author's nested-PCR control. The authors also applied the assay to whole blood and amniotic fluids although no data was presented. They also demonstrated the potential usefulness of the method to compare the efficacy of different treatment regimes and determine the prognostic value of treatment. Costa *et al.* (2000) described a LC real-time assay also based on the B1 gene which had a quantitative range of detection of between 0.75 to 0.75 x 10^6

parasites per PCR reaction. Application of the assay to four patients with toxoplasma reactivation demonstrated that parasite count could be determined and correlated with the clinical symptoms and treatment. Reischl *et al.* (2003a) compared the sensitivity and specificity of two LC real-time assays for the detection of *T. gondii*. One assay targeted the 25-fold repeated B1 gene and the other targeted a novel 529 bp DNA repeat. The 529 bp fragment appears to be repeated more than 300 times in the *T. gondii* genome providing a 10 to 100-fold higher sensitivity assay than targeting the B1 gene. This increased level of sensitivity was confirmed by application of both of the assays to 51 *T. gondii* positive human amniotic fluids. Real-time PCR has also been applied to the detection and quantitation of *T. gondii* oocysts in animal tissues as a means of screening raw food prior to human consumption (Jauregui *et al.*, 2001). Current methods to demonstrate the presence of *T. gondii* tissue cysts in food products require *in vivo* biological assays, which are laborious, costly and time consuming. A 5' nuclease assay was developed to target the 18S rRNA gene of *T. gondii* and was applied to the detection of cysts in tissues from experimentally infected mice and pigs as well as bradyzoite-spiked pig muscle samples. The results demonstrated that this assay may be useful to monitor *T. gondii* contamination in meat products and therefore may ultimately lead to a reduction in transmission of this parasite to human hosts.

Detection of Leismania Organisms by Real-Time PCR

Leishmaniasis is a serious disease which is endemic throughout tropical and subtropical regions. The incidence of new cases of infection is thought to exceed 1.5 million per year world-wide with at least 12 million people being affected and 350 million at risk. Leishmaniasis is caused by protozoa of the *Leishmania* genus and the disease can be generally classified into three major forms, cutaneous leismaniasis (CL), mucocutaneous leishmaniasis (MCL), and visceral leishmaniasis (VL) in increasing order of severity. CL can be caused by any pathogenic *Leishmania* strain, MCL is usually associated with strains belonging to the *Leishmania brasiliensis* complex and VL is generally caused by the *Leishmania donovani* complex. Serological assays for diagnosis show a high degree of cross-reactivity and cannot discriminate between current and previous infections. Diagnosis is therefore based on detection of parasites in clinical samples by histology or parasite culture. These methods are obviously labour intensive and time consuming. A real-time LC PCR assay was developed for the detection, quantitation, and discrimination of the three disease causing groups that targeted the 18S rRNA gene (Schulz *et al.*, 2003). The assay had a sensitivity suitable for the confirmation of VL from peripheral blood samples and was successfully used to detect parasites in a range of blood and tissue samples. Discrimination between the *L. donovani* complex the *L. brasiliensis* complex and other *Leishmania* strains was achieved through melting curve analysis. This was a significant improvement on the three previously published assays which could not differentiate strains into the three clinically relevant *Leishmania* groups (Bretagne *et al.*, 2001; Nicolas *et al.*, 2002; Wortmann *et al.*, 2001). Quantitative assays may prove useful for monitoring the efficacy of anti-leishmanial treatment however further studies are required to validate their utility.

Real-Time PCR Applications in Clinical Virology

Real-time PCR methods have proven to be useful tools in the diagnosis and management of a wide range of viral diseases (Niesters, 2002). Table 3 lists some of the real-time PCR assays that have been described which target a wide range of viral disease agents. The speed and specificity, plus their ability to directly quantify the target organism without post-PCR manipulations has made them the method of choice for a wide range of applications. Some of the areas in which real-time PCR methods have been applied in clinical virology will be reviewed below.

Detection of Respiratory Viruses by Real-Time PCR

Some of the viruses that commonly cause respiratory disease include influenza virus, parainfluenza virus, respiratory syncytial virus, adenovirus, and the more recently recognised human metapneumovirus. The recent emergence of the corona virus-associated severe acute

Table 3. Application of real-time PCR to the detection of clinically significant viruses.

Organism	Gene(s) targeted	Detection/ Quantitation	Sensitivity	Reference
Adenovirus	Hexon gene	Q	1.5×10^1 to 1.5×10^8 copies/PCR	Heim et al., 2003
Human Cytomegalovirus (HCMV)	major immediate-early gene	Q	10^1 to 10^7 copies/PCR	Nitsche et al., 2000
HCMV	US17 gene	Q	20 to 10^7 copies/PCR	Machida et al., 2000
HCMV	UL83 region	Q	10 to 10^6 copies/PCR	Gault et al., 2001
HCMV	UL83 gene	Q	10 to 10^4 copies/PCR	Griscelli et al., 2001
HCMV	glycoprotein B	Q	2×10^3 to 5×10^8 CMV DNA copies/ml blood	Kearns et al., 2001b
HCMV	DNA polymerase gene	Q	10^2 to 10^6 copies/PCR	Sanchez et al., 2002
HCMV	US17 gene	Q	500 to 50,000 CMV DNA copies/ml plasma	Stocher and Berg., 2002
HCMV	UL123 gene	Q	250 copies/ml plasma	Leruez-Ville et al., 2003
HCMV	immediate-early antigen gene	Q	not reported	Li et al., 2003
Dengue virus	conserved core regions	Q	10 to 10^7 PFU/ml	Shu et al., 2003
Dengue virus	N5S, capsid (C), UTR[a]	D	0.1 to 1.1 PFU detection limit	Callahan et al., 2001
Ebola, Marburg, Lassa, Crimean-Congo hemorrhagic Fever, Rift Valley Fever, Dengue, Yellow Fever Virus	L, GPC, NP, G2, 5' non-coding region, 3' non-coding region	Q	8.6 to 16 RNA copies/PCR	Drosten et al., 2002
Epstein-Barr Virus (EBV)	BALF5 gene	Q	2 to 10^7 copies EBV DNA/PCR	Kimura et al., 1999

Table 3 continued

EBV	BNRF1 p143 gene	Q	100 to 10^7 copies/ml plasma or serum	Niesters et al., 2000
EBV	EBNA1 gene	Q	10 copies/PCR	Stevens et al., 2002
EBV	BZLF1 gene	Q	10 to 10^9 copies/PCR	Patel et al., 2003
Enterovirus	5' non-coding region	D	11.8 enterovirus GE/PCR	Verstrepen et al., 2001
Enterovirus	5' non-coding region	D	50 enterovirus GE/PCR	Monpoeho et al., 2002
Enterovirus	5' non-coding region	D	not reported	Nijhuis et al., 2002
Enterovirus	5' non-coding region	D	$0.1 \text{ TCID}_{50}^{b}$	Rabenau et al., 2002
Enterovirus	5' non-coding region	D	510 copies/ml CSF	Lai et al., 2003
Hepatitis B virus (HBV)	S and X genes	Q	10^1 to 10^8 DNA copies/PCR	Abe et al., 1999
HBV	Pre S gene	Q	373 to 10^8 copies/PCR	Pas et al., 2000
Hepatitis C Virus (HCV)	5' non-coding region	Q	10^3 to 10^7 copies/PCR	Martell et al., 1999
HCV	5' non-coding region	Q	64 to 4,180,000 IU/ml serum	Kleiber et al., 2000
HCV	not reported	Q	10^3 to 10^7 RNA copies/PCR	Yang et al., 2002
Herpes Simplex Virus (HSV)	DNA polymerase	D	not reported	Espy et al., 2000b
HSV	DNA polymerase	D	12.5 copies/PCR	Kessler et al., 2000
HSV, varicella zoster virus (VZV), CMV	gD gene, gG gene, DNA polymerase, gene 38	D	580 copies/ml (HSV-1), 430 copies (HSV-2)	van Doornum et al., 2003
HSV	glycoprotein gene	Q	<10 to 10^8 copies/PCR	Ryncarz et al., 1999
HSV, VZV	gD gene, gG gene, DNA polymerase	D	<10 plasmid copies/PCR	Weidmann et al., 2003
Human Herpesvirus 6 (HHV-6)	U22 gene	Q	10 GE/PCR	Collot et al., 2002

Table 3 continued

HHV-6	U65-U66 genes	Q	10 GE/PCR	Gautheret-Dejean et al., 2002
HHV-6	U67 gene	Q	1 GE/PCR (1 to 10^6 GE/PCR)	Locatelli et al., 2000
HHV-6	large tegument protein gene	Q	10 GE/PCR	Kearns et al., 2001a
HHV-6, Human Herpesvirus 7 (HHV-7)	DNA polymerase gene	Q	10 GE/PCR	Safrontez et al., 2003
Human HHV-7	U100 gene	Q	14 GE/PCR	Fernandez et al. 2002
Human Herpesvirus (HHV-8)	Major capsid protein gene	Q	10 GE/PCR	Broccolo et al., 2002
HHV-8	K5, ORF[c]25, ORF37, ORF47, ORF56	Q	5-50 GE/PCR	Stamey et al., 2001
Human Metapneumovirus (hMPV)	N gene	D	not reported	Mackay et al., 2003
hMPV	N, M, F, P, and L genes	D	100 copies/PCR	Cote et al., 2003
Influenza viruses A and B	matrix protein (A) and hemagglutinin (B)	Q	13 copies/PCR (A) and 11 copies/PCR (B)	van Elden et al., 2001
Norovirus	ORF 1-ORF 2	Q	10 copies/PCR	Kageyama et al., 2003
Respiratory Syncytial virus (RSV)	N gene	Q	2.3×10^{-3} to 2.3×10^2 PFU	Hu et al., 2003
RSV	N gene	Q	not reported	Gueudin et al., 2003
Smallpox virus	hemagglutinin	D	5 to 10 GE/PCR	Espy et al., 2002
Smallpox virus	hemagglutinin	D	25 GE	Sofi Ibrahim et al., 2003
VZV	gene 28 and gene 29	D	not reported	Espy et al., 2000a
West Nile virus (WNV)	novel sequence	D	<1 PFU virus/PCR	Lanciotti et al., 2000

Notes:
[a] UTR: untranslated 3' region;   [b] TCID$_{50}$: 50% tissue culture infective dose;   [c] ORF: open reading frame.

respiratory syndrome (CDC SARS Investigative Team, 2003) has highlighted the need for rapid and specific diagnostic tests to differentiate between patients suffering from influenza and the other more common viral and bacterial respiratory pathogens and those infected with the SARS-associated coronavirus.

Influenza virus is a common respiratory pathogen that is highly contagious and responsible for considerable morbidity and mortality especially during the winter months. The elderly, the immunocompromised and patients with respiratory problems such as emphysema and asthma are particularly vulnerable to developing severe illness. Rapid detection of the virus may also be useful for timely initiation of antiviral therapy using new anti-influenza virus compounds such as neuraminidase inhibitors. Conventional detection methods for influenza virus include cell culture, shell vial culturing, antigen detection and serological tests. However, these tests are either too slow to allow timely diagnosis or they lack sensitivity and specificity. The first real-time PCR assay for the detection of influenza virus was recently described by van Elden *et al.* (2001). The assay simultaneously detects influenza virus A and B using a multiplex 5' nuclease format. Primers were designed to target the matrix protein of influenza A virus and the hemagglutinin gene segment of influenza B virus. The assays were demonstrated to have a sensitivity of 13 viral RNA copies and 11 viral RNA copies for influenza A virus and influenza B virus respectively. The assays were applied to the detection of influenza virus in 98 clinical samples from patients with upper or lower respiratory tract symptoms with the results being compared to conventional viral culture and shell vial culturing. The real-time PCR assay was found to be more sensitive than the conventional methods and follow-up of six symptomatic patients demonstrated that influenza virus could be detected for up to seven days after infection using the assay. During this time the patients were still symptomatic although the virus could no longer be isolated by culture after day one or two for the majority of the patients. Using serial dilutions of electron microscope counted stocks of influenza virus the authors demonstrated that the assay could be used in a quantitative format. The assay may be useful for determining viral load during antiviral therapy, however this needs to be confirmed by future studies.

Respiratory syncytial virus (RSV) is the most common etiological agent of lower respiratory tract disease in children. The virus has been classified into two subgroups, A and B based on antigenic and genetic variations in the structural proteins. Detection of RSV is based on virus culture, which is slow and relatively insensitive, or ELISA or immunofluoresence (IF) which also have limited sensitivity and specificity. Conventional RT-PCR assays have been described for the detection of RSV and have been demonstrated to be sensitive and specific diagnostic methods, however they do require post-PCR analysis. Hu *et al.* (2003) described a real-time QPCR assay for the detection, subgrouping and quantitation of RSV. The assays utilised two pairs of primers targeting the conserved nucleocapsid (N) gene specific for detection and subgrouping of the A and B virus subtypes. The assays were demonstrated to be sensitive, 0.023 plaque forming units (PFU) or two copies viral RNA, and 0.018 PFU or nine copies of the viral RNA for the RSV A and B subtypes respectively. The assays were applied to the detection of RSV in 175 NPA from children with respiratory disease with the results being compared to culture and IF. Overall the new assay was demonstrated to be about 40% more sensitive than culture and 10% more sensitive than IF for the detection of RSV in the samples in the study. In addition the new assay was able to subgroup the 36 RSV positive samples directly with 10 being identified as RSV A and 26 identified as RSV B. The assay also demonstrated a linear range of quantitation of between 2.3×10^{-3} to 2.3×10^2 PFU per reaction when applied to a virus stock sample. This QPCR assay may prove useful in studies to determine the relationship between viral load and subgroup with disease severity.

Gueudin *et al.* (2003) described a LC RT-PCR assay for the quantitative detection of RSV in nasal aspirates of children with respiratory disease. The assay was developed to investigate the relationship between disease severity and the amount of RSV in nasal aspirates. Similarly to Hu *et al.* (2003) they targeted the N gene but they also included a second control QPCR assay in the study specific for the glyceraldehyde-3-phosphate dehydrogenase (GAPDH) house-keeping gene. RSV particles are mainly cell-associated and nasal aspirate samples are non-homogenous therefore the quantitative GAPDH RT-PCR assay was used to standardise the quantitative results of the RSV assay.

RSV and GADPH were quantified in nasal aspirates from 75 children hospitalised for acute respiratory tract disease with the results being compared to culture and IF. The QPCR assay was more sensitive than the other two methods with 42 samples positive in the assay, IF detected RSV in 31 samples and culture detected RSV in 34 samples. The RSV RNA quantitation results for the RSV positive samples were then compared to the severity of disease in the patients. The mean number of RSV RNA copies was slightly higher in the severe disease group (4.05×10^7 copies) compared to the non-severe group (9.1×10^6 copies) with the amounts of RNA in individual samples being roughly related to severity of disease. However, when the GAPDH RNA quantitation results were taken into account as in indication of total cell number, the differences were no longer significant. The authors concluded that although the assay may not prove useful as a quantitative assay it is suitable as a qualitative diagnostic test for RSV infection.

Human metapneumovirus (hMPV) is a newly identified member of the *Paramyxoviridae* that has been isolated in NPA from patients with respiratory disease. The virus has been shown to cause a disease similar to RSV with patients displaying varying symptoms ranging from a wheeze to bronchiolitis. The virus is difficult to grow in cell culture therefore a rapid and sensitive diagnostic test is required to determine its role in respiratory disease in humans. Mackay *et al.* (2003) and Cote *et al.* (2003) both recently described real-time RT-PCR assays for the detection of hMPV. Mackay *et al.* (2003) developed a LC 5' nuclease assay targeting the N gene sequence of hMPV that they compared to a PCR-ELISA assay they had also developed. The analytical sensitivity of the two assays was the same, however the real-time PCR assay proved more sensitive when applied to the detection of hMPV in clinical samples. In the study of Cote *et al.* (2003) five sets of primers targeting the viral nucleoprotein (N), matrix (M), fusion (F), phosphoprotein (P), and polymerase (L) genes were compared in LC SYBR green quantitative RT-PCR assays. The assays were applied to 20 viral cultures with characteristics of hMPV cytopathic effect and the PCR positivity rates were 100, 90, 75, 60, and 55% using the N, L, M, P and F primers respectively. The five sets of primers were also evaluated for their ability to detect hMPV in 10 NPA samples. All of the ten samples were positive for hMPV using the N primers and the other primers detected hMPV in 6, 8, 3, 8 of the samples using the M, F, P, and L gene primers respectively. The authors concluded that the N primers were the most suitable for a diagnostic assay due to their superior sensitivity of approximately 100 copies per reaction. Both of the hMPV assays described in these two studies may prove useful in the detection of hMPV in samples from patients with respiratory tract infections, however they both require further validation through the application of them to larger numbers of clinical samples. These studies will also provide more accurate assessment of the prevalence of hMPV as a cause of human respiratory disease.

Detection of Herpes Viruses by Real-Time PCR

The human herpes virus family includes the clinically important pathogen herpes simplex viruses 1 and 2 (HSV-1 and HSV-2), varicella-zoster virus (VZV), and the human herpes viruses 6, 7, and 8 (HHV-6, HHV-7, HHV-8). HSV causes a wide spectrum of clinical disease with a variety of anatomical sites being infected including the skin, lips, oral cavity, eyes, genital tract and the CNS. Disseminated HSV infection is a serious condition that can occur in immunocompromised patients and in neonatal infection acquired by transmission of the virus through the infected birth canal of the mother. Although CNS infection by HSV can be fatal, effective therapeutic management of infections is possible however this requires rapid diagnosis to enable timely administration of antiviral therapy. Conventional methods for the detection of HSV are based on cell culture, which is the current gold standard for all specimens except CSF samples from patients with CNS infections. PCR has become the gold standard for diagnosing CNS infections, however conventional PCR methods were slow and lacked specificity. This has led to the development of a number of real-time PCR assays for the rapid detection of HSV in clinical samples. Kessler *et al.* (2000) described a LC 5' nuclease assay targeting the DNA polymerase gene of HSV, which was compared to an 'in-house' conventional PCR for the detection of HSV in 59 CSF specimens. The real-time assay and the conventional PCR assay were demonstrated to have very similar levels of sensitivity. A real-time QPCR assay for the detection of HSV was described by Ryncarz *et al.* (1999). The assay was demonstrated to have a linear range of detection of <10 to 10^8 copies of HSV DNA per reaction. The assay was applied to the detection of HSV in 335 genital tract specimens from HSV-2 seropositive patients and 380 CSF samples with the results being compared to tissue culture and an 'in-house' gel-based liquid hybridisation system. Overall the new assay was slightly less sensitive than the gel-based liquid hybridisation system.

VZV is the causative agent of chicken pox and shingles however in immunocompromised patients it can cause severe disease of the CNS or respiratory tract. Disseminated CNS infections can be fatal however rapid diagnosis allows timely initiation of antiviral therapy. Weidmann *et al.* (2003) developed a panel of 3 5' nuclease assays on the LC for the detection of HSV-1, HSV-2, and VZV in clinical samples. All three real-time assays demonstrated a high analytical LOD (<10 plasmid copies per assay) similar to the authors 'in-house' nested conventional PCR assays. The authors applied the new assays to the detection of HSV and VZV in 106 clinical samples including CSF, skin swabs, biopsies, vitreous body samples, blood samples and BAL. These were previously collected from immunocompromised patients, tested in the 'in-house' assays and then stored at -20^0C for up to 5 years before re-testing in the real-time assays. Results of the 'in-house' assay previously determined that 46 samples were positive for HSV-1, four were positive for HSV-2, and 56 were positive for VZV. In comparison to the 'in-house' assays, the sensitivity of the real-time assays for the detection of HSV-1, HSV-2, and VZV in the samples was 95%, 100%, and 96% respectively. A few samples, which had previously shown to be positive in the 'in-house' assays, were negative in the real-time assays however this may be due to template degradation during storage of the sample for up to 5 years prior to re-testing.

In a similar study, van Doornum *et al.* (2003) described a set of 5' nuclease assays for the detection of HSV-1, HSV-2, VZV, and human cytomegalovirus (HCMV). A universal control DNA, seal herpesvirus type 1 (PhHV-1) was added to each clinical sample prior to extraction and testing to monitor the DNA extraction and subsequent DNA amplification. Each of the viral targets was detected in separate assays because multiplexing the primers reduced the sensitivity of detection. The PhHV-1 was added at a low concentration to each clinical sample and was then extracted and detected to provide an internal control for inter-assay precision and reproducibility. This novel internal control system would be useful in other real-time PCR assays for microbial pathogens.

Detection of Enterovirus Infections by Real-Time PCR

Enteroviruses (EV) are common causes of CNS infections including viral meningitis and encephalitis. It has been estimated that they are responsible for 70-80% of cases of viral meningitis. The current gold standard for the diagnosis of enteroviral infections of the CNS is growth of the virus in cell culture. However, this test is slow requiring several days to produce results plus there can be difficulties propagating some EV types in cell culture. Rapid and reliable diagnosis of EV in CNS infections would provide a rational basis for antimicrobial therapy and limit the need for unnecessary procedures and irrelevant treatment. This has led to the development of alternative molecular based methods for the detection of EV in clinical specimens. The first assays to be described were conventional RT-PCR assays often using a nested format but more recently real-time assays have been developed for the rapid diagnosis of EV infections. Verstrepen *et al.* (2001) developed a 5' nuclease real-time QPCR assay utilising a single-tube RT and amplification format. The assay targeted the 5' non-coding region, which was also targeted by several other real-time assays, and was demonstrated to have a LOD of 11.8 EV GE per PCR reaction. The assay was applied to the detection of EV in 70 CSF specimens from patients with suspected viral meningitis, with 19 of the samples being positive in the assay and 17 being positive by culture. The assay of Monpoeho *et al.* (2002) which included an internal inhibition control was applied to the detection of EV in 104 CSF samples collected during an outbreak of EV meningitis with the results being compared to viral culture. The assay detected EV in 61 of the CSF samples however 41 of these were culture negative for EV. Using a 'consensus positive' for EV infection, defined as a culture-positive or RT-PCR positive result and clinical evidence of EV meningitis the sensitivity and negative predictive value of the assay was 96.8% and 95.3% respectively. Further application of real-time PCR assays for EV and other agents of viral meningitis to samples from patients with viral meningitis will validate their use as the primary diagnostic test for viral CNS infections. Introduction of rapid tests will hopefully improve the diagnosis of viral CNS infections and lead to improvements in the management and treatment of patients.

Detection of Viral Hemorrhagic Fever Viruses by Real-Time PCR

Viral hemorrhagic fever (VHF) is a clinical syndrome caused by a number of different viruses including Marburg virus (MBGV), Ebola virus (EBOV), Lassa virus (LASV), Junin, Machupo, Sabia, and Guanarito viruses, Rift Valley Fever virus (RVFV), Hanta viruses, yellow fever virus (YFV) and dengue virus (DENV). Clinical manifestations of VHF infections include diarrhoea, myalgia, cough, headache, pneumonia, encephalopathy, hepatitis and VHF can prove to be fatal. The most characteristic manifestation is haemorrhage although non-haemorhagic infections do occur. Definitive diagnosis is usually based on laboratory tests with rapid definitive diagnosis being important to identify the causative virus so specific treatment can be initiated and perhaps more importantly to exclude non-VHF infections. Rapid diagnosis is also important for case management such as the use of isolation procedures and the tracing of patient contacts. Conventional PCR assays have been described for all of the agents of VHF however many of them are relatively slow and require a separate RT step prior to amplification. Drosten et al. (2002) recently described a set of six one-step real-time PCR assays for the detection of MBGV/EBOV, LASV, CCHFV, RVFV, DENV, and YFV. The assays can be performed in two separate runs on the LC with a pair of different universal cycling conditions. The RVFV, DENV, and YFV assays were 5' nuclease assays and the MGBV/EBOV, LASV and CCHFV assays were SYBR Green I assays. The six assays were demonstrated to be specific for the individual agents and their analytical sensitivity using spiked negative human plasma was determined to be between 1,545 to 2,835 viral GE/ml serum (8.6 to 16 RNA copies per assay). The assays were applied to the detection of VHF agents in 30 samples from suspected cases of VHF with RNA extraction and

PCR detection taking less than 6 h. The amounts of VHF RNA in the samples were quantified using a quantitative standard run in parallel with the patient samples in the assay. The patients were found to have high viral DNA loads in their serum with virus DNA concentrations in acute phase patients being orders of magnitude above the detection limit of the assays indicating that the assays were sufficiently sensitive to diagnose VHF during acute phase infection.

Additional real-time PCR assays targeting DENV the causative agent of dengue fever, have also been described. Shu *et al.* (2003) developed group and serotype-specific one-step SYBR green LC assays targeting conserved sequences in the core region of the DENV genome. The assays were demonstrated to detect, differentiate and quantify all four different DENV serotypes in acute phase serum. The assays had a dynamic range of detection of 10 to 10^7 PFU/ml for cell-culture derived DENV and a detection limit of between 4.1 and 10 PFU/ml depending on the serotype. The assays were applied to the detection of DENV in 193 acute phase serum samples from confirmed dengue fever patients with the real-time assays appearing to be more sensitive than the conventional cell culture method. In a similar study, Callahan *et al.* (2001) developed 5' nuclease assays for the detection and differentiation of serotypes 1 to 4 and a group-specific assay for the detection of DENV. The assays were applied to 67 dengue viraemic human sera and demonstrated high sensitivity and specificity.

The suspicion of a VHF infection in a patient immediately raises a number of questions which require answers as soon as possible. Most importantly, the infectious agent must be determined so its potential for further spread can be assessed and appropriate control measures and contact tracing can be instigated. Real-time PCR methods may prove very useful to determine the infectious agent in such circumstances.

Viral Genome Quantitation by Real-time PCR

Viral genome quantitation has become increasingly useful in the diagnosis and management of viral disease. For example, viral quantitation or viral load (VL) testing by real-time PCR has proven useful as a diagnostic marker in human cytomegalovirus (HCMV) infection with such testing aiding the diagnosis of active disease. It is also a useful prognostic marker in infections such as HIV for assessing disease progression and as a therapeutic marker for monitoring the efficacy of antiviral chemotherapy in patients with chronic HIV, hepatitis B or hepatitis C infection. The measurement of VL has proven useful in assessing an individuals' infectivity, for example to estimate the risk of vertical transmission of HIV between mother and child. The role of VL testing by PCR in clinical virology has recently been reviewed by Niesters (2001), Berger and Preiser (2002) and Mackay *et al*. 2002. The first generation of QPCR methods to be described used end-point detection utilising internal or external controls of known concentrations, which were amplified in parallel with the samples of interest. Following PCR cycling the unknown test samples were compared to the control values and a quantitative value assigned to the test samples. Because the final amount of accumulated product at the end of the PCR process is very susceptible to minor variations in reagents and sample matrices, there are limitations on the accuracy of QPCR methods based on end-point detection. Real-time PCR methods provide much more accurate quantitation because reactions are quantified by the point in time during cycling when amplification is detected rather than the amount of PCR product accumulated after a fixed number of cycles. Using fluorescent monitoring of the PCR reaction, the increase in specific amplification product can be monitored in real-time, which allows the accurate quantification over six orders of magnitude of the DNA or RNA target sequence.

Quantitation of HCMV by real-time QPCR has become a very useful diagnostic test for the management of immunocompromised patients following solid organ or bone marrow transplantation (BMT), in individuals with HIV infection and in other patients on immuno-suppressive therapy. HCMV can cause a severe disease in such individuals and accurate quantitation methods have demonstrated that patients with high viral load are at greater risk of developing disease. In addition VL testing is useful for monitoring the efficacy of antiviral therapy. Various genes have been targeted in these assays. Nitsche *et al*. (2000) compared ten different primer and probe combinations targeting the major immediate-early gene locus in a 5' nuclease assay for HCMV quantitation, the results being compared to the pp65 antigen detection assay. Sensitive detection over 6 orders of magnitude was demonstrated with the HCMV DNA quantitation results correlating with the pp65 antigen assay in plasma samples from bone marrow transplant (BMT) patients. The US17 gene was targeted in the 5' nuclease QPCR assay of Machida *et al.* (2000) in a pilot study to detect HCMV in BMT patients however only one patient was demonstrated to have definite HCMV disease. Kearns *et al.* (2001b) compared a LC QPCR assay targeting the glycoprotein B gene with a 5' nuclease quantitative HCMV assay previously validated in a comparison with the pp65 antigenemia assay for the detection of HCMV in whole blood extracts from immunocompromised patients. The viral load ranges correlated between the two assays with a LOD of approximately 10 copies and a linear range of quantitation of between 2×10^3 to 5×10^8 copies. The authors also successfully applied the LC QPCR assay to the detection of HCMV in urine and respiratory samples (Kearns *et al.*, 2001c; Kearns *et al.*, 2002b).

Griscelli *et al.* (2001) described a 5' nuclease QPCR assay that targeted the UL83 gene of HCMV and facilitated normalisation of the quantitative data through the detection and quantitation of a human house-keeping gene. The UL83 gene encoding the pp65 antigen copy number was quantified and the GAPDH gene copy number was used to normalise the data. A QS containing the both the UL83 and GAPDH genes was used to generate a standard curve with the assay demonstrating a linear range of detection of 10 to 10^4 copies per PCR. The QPCR assay was shown to be more sensitive than the pp65 antigen assay in BMT patients and has the potential to be multiplexed into a single-tube assay. A multiplex 5' nuclease QPCR for the detection of the HCMV DNA polymerase gene with normalisation of the data using a human apoprotein B (HAPB) gene assay was also described (Sanchez *et al.*, 2002). The assay was applied to the detection and quantitation of HCMV in plasma from adult lung transplant patients. The second PCR also acts as an internal control against PCR inhibition, which can lead to false-negatives. All of these assays provided quantitative data, based on the assumption that the amplification efficiencies of the QS and samples were equal which is not always the case. The use of internal controls (IC) in a LC QPCR assay for HCMV was investigated by Stocher and Berg (2002). In their LC assays they spiked patient samples with a known amount of heterologous competitor DNA and used an algorithm to normalise the data due to the varying amplification efficiencies between the QS and samples. The IC was a plasmid construct that could be co-amplified with the same HCMV primers but contained the neomycin phosphotransferase gene. This method successfully controlled for PCR inhibition in the patient samples and is a model for the design of other real-time QPCR assays for the detection of other microbial targets. In these studies HCMV VL in various blood compartments such as peripheral blood leukocytes (PBL) and plasma were investigated however the usefulness of VL detection in whole blood (WB) in comparison with PBL was not reported. Using a LC assay, Mengelle et al. (2003) demonstrated that automated DNA extraction and quantitation of HCMV DNA from WB samples provided acceptable results when compared to manual extraction and detection. The authors also demonstrated good correlation between viral loads in WB and PBL leading them to conclude that the use of WB instead of separated, counted PBL can be used to monitor HCMV-infected patients. Li et al. (2003) determined HCMV loads in WB samples from transplant patients and demonstrated that the loads correlated with those determined using the antigenemia assay. They demonstrated that antigenemia values of 1 to 2, 10, and 50 positive cells per 2×10^5 leukocytes correlated with HCMV VL's of 1,000, 4,000, and 10,000 copies/ml respectively, and proposed these as cut-off points for initiating antiviral therapy in patient groups with high, intermediate, and low-risk of HCMV diseases.

Epstein-Barr virus (EBV) infects more than 90% of the population world-wide and in immunocompetent individuals the virus establishes a life-long asymptomatic infection. In a minority of immunocompetent individuals the virus causes infectious mononucleosis and EBV-related malignancies such as Burkitt's lymphoma and nasopharyngeal carcinoma. In immunocompromised patients, active EBV infection is a strong risk factor for the development of post-transplant lymphoproliferative disease (PTLD), AIDS-related lymphoma, and X-linked proliferative syndrome. The monitoring of EBV load in transplant patients has been shown to be a useful tool in the diagnosis and management of PTLD. A number of real-time QPCR assays for

EBV VL testing have been described. Kimura *et al.* (1999) developed a 5' nuclease QPCR assay targeting the BALF5 gene encoding the viral DNA polymerase that had a linear range of 2 to 10^7 copies EBV DNA. The assay was used to measure the EBV VL in peripheral blood mononuclear cells (PBMNC) in patients with symptomatic EBV infection. The authors established the usefulness of the assay for diagnosing symptomatic EBV infection and proposed a diagnostic cut-off value of $10^{2.5}$ copies/µg PBMNC DNA (for symptomatic EBV disease). In this study all patients with PTLD and chronic active EBV infection had EBV VLs much greater than this cut-off value. Niesters *et al.* (2000) applied a 5' nuclease QPCR assay targeting the BNRF1 p143 gene encoding the non-glycosylated membrane protein, to the quantitation of EBV in transplant patients. Plasma and serum samples were screened over a range of between 100 and 10^7 copies of DNA per ml using two sample preparation methods. EBV DNA was detected in 19.2% of immunosuppressed solid-organ transplant patients without symptoms of EBV disease with a mean VL of 440 copies/ml. EBV DNA was also detected in all transplant patients with PTLD with a mean VL of 544,570 copies/ml and no EBV DNA was detected in healthy individuals in non-immunocompromised control groups and a mean VL of 6,400 copies/ml was detectable in patients with infectious mononucleosis. The detection of latent EBV in the immunosuppressed patients without symptoms of EBV disease is in keeping with previous reports suggesting the presence of EBV DNA alone does not always indicate active EBV disease. Two LC QPCR assays have also been described for EBV VL testing, one targeting the EBNA1 gene (Stevens *et al.*, 2002) and one targeting the BZLF1 gene (Patel *et al.*, 2003). These assays were also demonstrated to be useful for the diagnosis of active EBV disease. Although these QPCR assays have proven useful in the diagnosis and management of patients with active EBV disease, inter-laboratory standardisation of these methods has not yet been achieved. Future prospective studies investigating the EBV VL in immunosuppressed patients using these assays will help to establish clinically relevant cut-off values, preferred sample types and extraction methods, and standardised QS and ICs.

Approximately 300 million people world-wide are chronically infected with Hepatitis B virus (HBV). The measurement of HBV DNA in patient serum has become a very useful tool in the management of HBV infection. HBV VL testing is useful to assess treatment response in patients undergoing antiviral therapy and can also be a marker for the emergence of resistant viral strains in patients undergoing lamivudine or famciclovir treatment. HBV VL testing is also useful to assess the infectivity of HBV carriers, for example to estimate the risk of vertical transmission from female HBV carriers to their infants. Conventional QPCR assays were developed for the quantitation of HBV in patients including several commercially produced tests. Although these commercially produced assays have proven to be useful many have a limited linear range of quantitation. The development of real-time PCR platforms has led to the development of a number of more rapid, real-time QPCR assays for HBV VL testing. Abe *et al.* (1999) described three 5' nuclease assays, two targeting the HBV surface gene (S gene) and one targeting the X gene. The assays were sensitive (LOD = 10 copies/PCR) and had a linear range of quantitation of 10^1 to 10^8 DNA copies per PCR. The assays were applied to 46 serum samples from patients with chronic HBV disease and the results compared to a commercially produced signal amplification assay with enzymatically labelled branched DNA (bDNA). The results between the two assays correlated however the real-time assay was more sensitive with eight of the 46 samples only being positive in this assay and not in the signal amplification assay. These eight samples were demonstrated to have HBV DNA levels below the detection limit of the bDNA assay with the real-time QPCR assay being approximately 10^4 to 10^5 times more sensitive. The real-time QPCR assay is a sensitive assay with a wide linear range of quantitation however; the usefulness of this assay as a tool for the management of chronically infected HBV patients has yet to be established. The real-time QPCR assay of Pas *et al.* (2000) targeted the pre-S gene of HBV and was demonstrated to have a dynamic range of detection of 373 to 10^{10} genome copies per ml. Application of the assay to the monitoring of four patients during antiviral treatment demonstrated the results of the new assay correlated closely with those from two commercial assays, the HBV Digene Hybrid Capture II Microplate assay and the Roche HBV MONITOR assay. All of the assays detected a rapid decrease in HBV VL in the four patients following initiation of lamivudine treatment. To obtain the same dynamic range of detection demonstrated by the real-time QPCR assay, some of the samples had to be re-tested in the commercial assays following a dilution or concentration step. This wide range of quantitation is a clear advantage of the real-time assay plus the likelihood of false-positives through amplicon carryover is also eliminated.

The RNA virus, Hepatitis C (HCV) is the most common cause of non-A and non-B viral hepatitis. Most infection is asymptomatic however in approximately 70-80% of patients, the infection becomes chronic. Chronic HCV infection can lead to a range of clinical outcomes from mild non-progressive liver damage, to severe chronic hepatitis that can progress to cirrhosis, end-stage liver disease, and hepatocellular carcinoma. Chronic infection is characterised by persistent viraemia, which is the standard diagnostic marker for chronic infection. HCV VL testing has only a limited value for determining disease prognosis, however it is a valuable tool for monitoring the success of antiviral therapy. HCV VL testing is also useful for estimating infectivity and determining the risk of mother-to-infant transmission.

Because HCV is an RNA virus HCV VL testing requires a reverse transcriptase step prior to PCR amplification and detection. Martell *et al.* (1999) described a 5' nuclease single-tube QPCR assay targeting the 5' non-coding region of the HCV genome. The assay demonstrated a linear 5-log range of detection with very low intra-assay and inter-assay coefficients of variation. Application of the assay to 79 RNA samples from the sera of infected patients revealed a correlation between the real-time assay and the Quantiplex 2.0 bDNA assay and the Superquant assay. A second real-time assay was described by Kleiber *et al.* (2000) which also targeted the 5' non-coding region of the HCV genome with the assay including an IC for inhibition. All HCV genotypes were amplified with equal efficiency and the assay had a dynamic range of quantitation of 64 to 4,180,000 IU/ml and a LOD of 40 IU/ml. The application of the assay to interferon-treated patients distinguished responders from non-responders and responder-relapsers. Both of these assays may prove useful in monitoring the efficacy and response to antiviral treatment in patients with chronic HCV infection.

There are six species of human adenovirus (HAdV) with 51 types associated with a variety of diseases affecting all organ systems. Recently, studies have shown that HAdV can cause serious life-threatening infections in immunocompromised hosts such as hematopetic stem cell transplant patients. These HAdV infections are frequently associated with a viraemia with one or more organ systems being involved producing symptoms including pneumonia, meningitis, hepatitis, rash, diarrhoea, or cystitis. The gold standard for the detection of HAdV is virus isolation however, that is slow and can require up to three weeks for cytopathic effects to develop. This led to the development of PCR-based diagnostic tests for the detection of HAdV in clinical specimens. However, qualitative detection of HAdV DNA in immunocompromised patients has a low predictive value because HAdV DNA can be detected occasionally in the blood of healthy, persistently infected individuals. Quantitation of HAdV DNA in blood samples may provide more informative data for predicating the risk of disseminated infection in immunocompromised patients. Heim *et al.* (2003) have recently described a 5' nuclease LC assay for the sensitive and specific detection and quantitation of all 51 HAdV types in blood samples from immunocompromised patients. The assay targets a conserved region of the hexon gene and has a sensitivity of approximately 15 copies per reaction with a linear range of quantitation of 1.5×10^1 to 1.5×10^8 per reaction. Application of the assay to 234 clinical samples including blood, serum, eye swabs, and faeces demonstrated the LC assay to be more sensitive than a previously developed conventional in-house PCR assay. HAdV viremia was detected in four of 27 pediatric and eight of 93 adult stem cell transplant recipients but only in five of 306 healthy controls. The HAdV VLs were greater in the pediatric patients when compared to the adult patients and the controls with a small number of immunocompromised children having very high VLs (up to 1.1×10^{10} copies/ml) which were associated with symptoms of disseminated disease. The QPCR assay is a useful tool in the management of immunocompromised patients at risk of HAdV infection and may provide more simplified and rapid diagnosis of disseminated disease. Although there is no specific antiviral therapy for disseminated HAdV infection, cidofovir and ribavirin have been used to treat severe infections in patients but no controlled clinical trial has been carried out. This QPCR assay may prove useful for monitoring HAdV VL in patients during clinical trials of antiviral agents for the treatment of disseminated infection.

Real-Time PCR assays for the Detection of Antibiotic Resistance, Antiviral Susceptibility and Toxin Genes

In addition to pathogen detection assays, real-time PCR methods have been applied to the detection of antibiotic resistance determinants and toxin genes in microbial pathogens. Some examples of real-time PCR assays for the detection of these gene targets are presented in Table 4. Assays for the detection of ciprofloxacin resistance have been described for *C. jejuni* (Wilson *et al.*, 2000), *Salmonella enterica* serotype *Typhimurium* DT104 (MR DT104) (Walker *et al.*, 2001), and *Yersinia pestis* (Lindler *et al.*, 2001). Resistance to quinolones such as ciprofloxacin is chromosomally mediated with mutations within the quinolone resistance-determining region (QRDR) of the DNA gyrase gene (*gyrA*) playing a major role in gram-negative bacteria. A LC assay with three probes was developed for the detection and differentiation of the Asp-87-to-Asn, the Asp-87-to-Gly, and the Ser-83-to-Phe mutations within MR DT104. Strains homologous to the probes could be distinguished from strains that had different mutations by their probe-target melting temperatures. Application of the assay to 92 isolates of MR DT104 demonstrated that 86 of the isolates possessed one of the three mutations described above. Although the method could not determine the *gyrA* mutation for six strains it did provide much more rapid results for the other 86 isolates, avoiding the need to sequence the *gyrA* to determine their mutation types.

A 5' nuclease assay was described for the discrimination of ciprofloxacin sensitive and ciprofloxacin resistant strains of *C. jejuni* (Wilson *et al.*, 2000). The assay targeted the C to T transition in codon 86 in the QRDR of *gyrA* and was able to rapidly and reliably differentiate between ciprofloxacin sensitive and ciprofloxacin resistant strains of *C. jejuni*. In the study of Lindler *et al.* (2001) ciprofloxacin resistance in *Yersinia pestis* was demonstrated to be due to a single nucleotide mutation in the QRDR of *gyrA*, with substitutions belonging to one

Table 4. Real-time PCR assays for the detection of antibiotic suseptibility, antibiotic resistance, antiviral susceptibility and toxin genes in clinical microbiology.

Organism	Function of assay	Gene(s) targeted	Reference
Bordetella pertussis	Pertussis toxin gene variants	ptxS1A, ptxS1B, ptxS1D, ptxS1E	Makinen *et al.*, 2002
Campylobacter jejuni	Ciprofloxacin resistance	*gyrA*	Wilson *et al.*, 2000
Corynebacterium diptheriae	Toxin gene detection	A and B subunit toxin genes	Mothershed *et al.*, 2002
Coxiella burnetti	Antibiotic suseptibility testing	com1 gene	Brennan and Samuel, 2003
Enterococcus faecalis, Enterococcus faecium	Oxazolidinione resistance	G2576U rRNA mutation	Woodford *et al.*, 2002
Helicobacter pylori	Clarithromycin resistance	23S rRNA	Oleastro *et al.*, 2003
H. pylori	Clarithromycin resistance	23S rRNA	Matsumura *et al.*, 2001
H. pylori	Clarithromycin resistance	23S rRNA	Gibson *et al.*, 1999
Mycobacterium tuberculosis	Rifampicin and isoniazid resistance	*rpoB* and *katG*	Garcia de Viedma *et al.*, 2002
M. tuberculosis	Rifampicin and isoniazid resistance	*rpoB* and *katG*	Torres *et al.*, 2000
M. tuberculosis	Rifampicin resistance	*rpoB*	Edwards *et al.*, 2001b
Staphylococcus species	Methicillin resistance	*mecA*	Killgore *et al.*, 2000
Salmonella enterica serotype Typhimurium DT104	Ciprofloxacin resistance	*gyrA*	Walker *et al.*, 2001
Streptococcus pneumoniae	Penicillin suseptibility	ptp2b gene	Kearns *et al.*, 2002a
Staphylococcus aureus	Fluoroquinolone resistance	*grlA*	Lapierre *et al.*, 2003
Vancomycin resistant enterococcus (VRE)	Vancomycin resistance	*vanA* and *vanB*	Palladino *et al.*, 2003a
VRE	Vancomycin resistance	*vanA* and *vanB*	Palladino *et al.*, 2003b
Yersinia pestis	Ciprofloxacin resistance	*gyrA*	Lindler *et al.*, 2001

of four groups. A LC assay was designed to differentiate between all four point mutations within the *gyrA* gene that resulted in ciprofloxacin resistance. The assay successfully discriminated between sensitive and resistant strains of *Y. pestis* on the basis of probe melting curve analysis following PCR amplification.

Clarithromycin resistance in *Helicobacter pylori* is associated with a single base mutation within the peptidyltransferase-encoding region of the 23S rRNA gene. The three mutations lead to the adenine residues at positions 2143 and 2144 being replaced with guanine (A2143G and A2144G) or cytosine (A2143C). Gibson *et al.* (1999) described a LC assay which could discriminate between clarithromycin sensitive and clarithromycin resistant strains using melting curve analysis of a probe complementary to the clarithromycin sensitive 23S rRNA sequence. In a similar study a hybridisation probe LC assay was described for the detection of mutations within the 23S rRNA gene conferring resistance to clarithromycin (Matsumura *et al.*, 2001). The authors used three probes and melting curve analysis to successfully identify clarithromycin resistant strains directly in gastric biopsy samples. Similarly, in the study of Oleastro *et al.* (2003) a hybridisation probe LC assay was designed to target the three point mutations (A2142C, A2142G, and A2143G) to determine clarithromycin susceptibility directly in gastric biopsy samples from patients with *H. pylori* infection. The assay differentiated between strains with the transition A to G in position 2142 or 2143, the transversion A to C in position 2142, and the wild-type sensitive genotype based on melting curve analysis. The assay was also shown to be able to detect resistance strains in mixed populations of *H. pylori* among sensitive wild-type strains. Rapid detection of resistance profiles of *H. pylori* directly in gastric biopsy specimens may prove useful in determining appropriate treatment regimens for the successful eradication of *H. pylori* from the patient.

The recommended treatment for *M. tuberculosis* is a combination of four drugs; rifampicin, isoniazid, pyrazinamide, and ethambutol with or without streptomycin. Multi-drug resistant (MDR) strains have arisen, most of which are resistant to rifampicin and at least one other drug. Detection of drug resistance in *M. tuberculosis* is usually performed by phenotypic testing, however this requires considerable time to produce a result. Early detection of drug resistance in strains is important to ensure the timely initiation of effective treatment regimens and to help reduce the risk of transmission of MDR strains. The molecular basis of drug resistance is becoming better understood with more than 95% of rifampicin resistant strains being associated with mutations within an 81 bp region of the *rpoB* gene. Resistance to isoniazid has also been reported with between 60 and 70% of isoniazid resistant strains having mutations within the *katG* gene, with a specific mutation in codon 315 being responsible for many cases of resistance. To facilitate the rapid detection of rifampicin and isoniazid resistance in *M. tuberculosis,* a number of real-time assays have been developed. Edwards *et al.* (2001b) reported a biprobe LC assay which detected all of the mutations present in the *rpoB* gene of 46 rifampicin resistant isolates of *M. tuberculosis.* A single-tube multiplex real-time LC assay for the detection of multiple rifampicin resistance mutations and high-level isoniazid resistance mutations in *M. tuberculosis* was reported by Garcia de Viedma *et al.* (2002). The assay was demonstrated to be able to detect 12 different mutations in eight codons within the whole *rpoB* core region, plus the most frequently occurring isoniazid resistance mutations.

Several other real-time PCR assays have been described for the detection of antibiotic resistance genes within a number of microbial pathogens. These include assays for the detection of the methicillin resistance gene, *mecA* the most frequent cause of methicillin resistance in staphylococci (Killgore *et al.,* 2000). The *vanA* and *vanB* genes of vancomycin resistant enterococci were targeted in the assays of Palladino *et al.* (2003a, 2003b). A real-time assay for the detection of fluoroquinolone resistance associated mutations in the *grlA* gene of *Staphylococcus aureus* was described by Lapierre *et al.* (2003). The assay demonstrated good correlation (98.8%) with the MICs of ciprofloxacin, levofloxacin and gatifloxacin when applied to 85 *S. aureus* isolates with varying levels of fluoroquinolone resistance.

Real-time PCR assays have also been developed for the rapid detection of bacterial toxin genes. *Corynebacterium diphtheriae* is the etiological agent of diphtheria. Laboratory confirmation of diphtheria relies on isolation of the organism from the patient and subsequent determination of the toxigenic potential of the isolate using the Elek test. Viable organisms are not always detectable in the specimens or are below the detection limit of culture. This led to the development of PCR-based detection tests, which are often the only laboratory test available for the diagnosis of diphtheria in the absence of culturable organisms. A 5' nuclease real-time PCR assay for the detection and differentiation of the A and B subunits of the toxin gene of *C. diphtheriae* was recently described (Mothershed *et al.*, 2002). The assays were demonstrated to be 750 times more sensitive than the conventional standard PCR assay currently considered the gold standard for molecular diagnosis of diphtheriae (LOD of 2 CFU compared to 1,500 CFU/reaction). The assay had a 100% sensitivity and specificity when applied to 23 toxigenic strains of *C. diphtheriae*, nine non-toxigenic strains of *C. diphtheriae*, and 44 other strains representing a diverse range of other respiratory flora and respiratory pathogens. The assay was also applied to the direct detection of toxigenic *C. diphtheriae* in 36 clinical specimens from patients with confirmed clinical diphtheria (35 throat swabs and one throat pseudomembrane). The real-time assay detected one or both of the subunits of the toxin gene in 34 of the 36 specimens compared to only nine specimens being found to be positive in the conventional PCR. This assay provides a more sensitive and rapid alternative to conventional PCR for the detection of toxigenic strains of this important pathogen.

Detection of Potential Bioterrorist Agents by Real-Time PCR

The deliberate release of *Bacillus anthracis* in the United States in 2001 has reaffirmed the need to develop rapid, real-time PCR assays for the detection of potential bioterrorism agents. *B. anthracis* causes a serious and often fatal disease of both humans and livestock making it a very effective biological weapon of mass destruction. Therefore rapid methods are required to detect this organism both in human patients and the environment. Pathogenic *B. anthracis* strains have a pair of plasmids, pX01 and pX02 which encode various virulence factors including the protective antigen, the capsule and the lethal factor. The first real-time PCR assay described for *B. anthracis* targeted the *cap* gene on the pX02 plasmid and utilised the Idaho LC and a SYBR gold detection format (Lee *at al.*, 1999). This work was one of the first to demonstrate the usefulness of the real-time detection format for anthrax detection and led to a number of subsequent studies. Qi *et al.* (2001) described a LC hybridisation probe assay which targeted the chromosomal *rpoB* gene which codes for the β-subunit of RNA polymerase. The specificity of the assay was confirmed by applying it to 144 *B. anthracis* strains, however it did not have the ability to determine if the strain detected was virulent or not. Many of the reported studies have concentrated on detecting anthrax spores in various types of environmental and clinical samples. Drago *et al.* (2002) utilised the LC assay of Qi *et al.* (2001) to develop a protocol for the rapid detection of *B. anthracis* in spiked nasopharyngeal specimens and wound specimens. The spiking studies demonstrated that the assay has the capability of detecting as few as 25 spores/ml diluent in about 90 min including the DNA extraction. Oggioni *et al.* (2002) also described a protocol for the rapid detection of anthrax spores in nasal swabs following enrichment. The assay was a multiplex LC assay targeting the lethal factor (*lef*) gene on pX01 and the chromosomal marker *rpo*B. With purified DNA the LOD of the assay was demonstrated to be about 50 genome equivalents but the detection level fell to 2,000 cell equivalents when spores were used as template in the assay. This level of detection would require swabs contaminated with between 60,000 and 600,000 spores/swab to ensure consistent detection, a scenario unlikely in a deliberate aerosol release. Therefore the authors recommended an overnight incubation to biologically amplify the numbers of cells present to detectable levels prior to PCR testing.

Makino *et al.* (2001) described a LC assay targeting the *cap* gene on pX01 which they used to detect anthrax spores from air samples. They filtered 100 litres of air through an air-monitoring device and then spiked the membrane filter with a single anthrax spore. The single spore was detectable in the real-time assay leading the authors to conclude that the assay may detect a single anthrax spore in air. However, its ability to detect such small numbers of spores in contaminated air samples was not validated. Bell *et al.* (2002) reported a LC real-time PCR assay for the detection of *B. anthracis* which targeted the protective antigen gene (*pagA*) and the capsulation B protein gene (*capB*) located on the pX01 and pX02 plasmids respectively. The assay was a hybridisation assay using melting curve analysis to confirm the product identity and is now produced by Roche Diagnostics as the LightCycler *Bacillus anthracis* detection kit. The assay targets genes on both of the plasmids allowing discrimination between virulent strains with both plasmids and non-virulent strains missing one or both plasmids. This strategy may prove useful if a non-virulent strain (lacking one or both plasmids) is deliberately released as a hoax agent. The assay was applied to 32 *B. anthracis* strains, 29 of which were considered to be virulent. Twenty-eight of the strains were positive for both *pagA* and *capA* genes however one isolate was negative in the *capA* assay. This isolate was later shown to have lost all of its encapsulation genes (*capA*, *capB*, and *capC*) implying that plasmid pX02 was lost also. Overall, the study demonstrated that the assay was a suitable method for the rapid identification of cultured isolates of *B. anthracis*. The assay was also demonstrated to have a level of sensitivity of 1 copy per µl of sample and therefore may prove useful for detecting *B. anthracis* in clinical samples however this needs to be validated by further studies.

Ellerbrook *et al.* (2002) developed a real-time assay that targeted both of the *pagA* and *capC* genes plus the chromosomal *rpoB* gene. The author's strategy to target genes on the chromosome and both plasmids was designed to enable differentiation between virulent and avirulent strains, which would be particularly useful in the event of a deliberate release of anthrax. The pX02 plasmid has been transferred into other *Bacillus* species and genes from the pX01 plasmid have been expressed in other bacteria therefore the authors concluded that inclusion of the chromosomal gene target in the assay is the only method to definitively identify pathogenic *B. anthracis* strains by PCR. The specificity of the assay was validated by applying it to a range of *B. anthracis* strains and other *Bacillus* species. Four of the five *B. cereus* strains gave a positive result in the *rpoB* gene assay however the C_T values were much higher suggesting that PCR mis-priming may have been responsible for these false-positives. The potential for detecting false-positives may limit the use of the assay as a method to definitively identify *B. anthracis* and perhaps alternative chromosomal gene targets should be investigated which may provide a greater specificity of detection.

The deliberate release of anthrax in the United States in 2001 provided the first opportunity to evaluate real-time PCR assays for anthrax with field samples. Workers at the CDC in Atlanta, USA applied real-time assays to confirm the identity of 317 isolates of *B. anthracis* obtained during investigations into the 2001 deliberate release. The assays targeted genes on the pX01 and pX02 plasmids plus a region of the *B. anthracis* chromosome however the precise gene targets were not identified and the primer sequences were not published. The assay demonstrated 100% specificity and sensitivity and are now used within the Laboratory Response Network (LRN) which includes over 200 laboratories in several countries and all 50 states of the USA as their diagnostic test for *B. anthracis*.

Y. pestis is the causative agent of plague, a serious disease that has caused some of the worst epidemics in human history. The transmission of plague occurs through the bite of infected fleas although human to human transmission can occur through respiratory droplets from cases of plague pneumonia. Diagnosis of plague requires culture of *Y. pestis* from clinical samples, however definitive identification can prove difficult with commercial biochemical identification systems. PCR methods may provide a rapid and more specific test for identification of suspect colonies following culture. The first real-time PCR assay for *Y. pestis* was described by Higgins *et al.* (1998) who developed a 5' nuclease assay which targeted the plasminogen activator factor (*pla*) gene. The assay was applied to the detection and quantitation of *Y. pestis* bacilli in experimentally infected fleas. However the specificity of the assay was compromised with several bacteria other than *Yersinia* species producing positive results in the assay. The majority of *Y. pestis* strains carry three virulence plasmids; the *Y. pestis*-specific pPla which encodes the *pla*; the pCD1 plasmid encoding several virulence factors referred to as YOPs (*Yersinia* outer membrane proteins) which are present in all pathogenic members of the species including *Y. pestis*, *Yersinia enterocolitica* and *Yersinia pseudotuberculosis*; and the pMT1 plasmid which encodes the highly immunogenic fraction 1 capsule antigen (F1) (*caf1*) and the *Y. pestis* murine toxin (*ymt*). Tomaso *et al.* (2003) developed a dual multiplex LC assay for the detection of the plasmid-encoded genes *pla*, *caf1* and *ymt* and the chromosomal 16S rRNA gene. These genes were previously shown to have copy numbers of 186, 4, 2 and 6 respectively, therefore targeting these genes may provide assays with a greater level of sensitivity than those targeting single copy genes. An internal amplification control was also included which detected bacteriophage λ DNA. The specificity of the assay was validated with a range of *Yersinia* species and other unrelated genera with *Y. pestis* strains only yielding positive results in the assay. The lower limit of detection was approximately 0.1 genome equivalent with rat and flea DNA having no inhibitory effects on the detection of *Y. pestis*. This assay may be a useful tool for screening suspect colonies to provide a rapid identification of *Y. pestis*. The high sensitivity of the assay may also make it useful for directly detecting *Y. pestis* in patient samples although this remains to be proven.

Francisella tularensis is the causative agent of tularemia a serious disease of humans. There are six recognised forms of tularemia; ulceroglandular, oropharyngeal, oculoglandular, pneumonic, septicaemic, and typhoidal. Untreated the overall mortality of tularemia is 4% for ulceroglandular and 30-50% for typhoidal, septicaemic and pneumonic types, however with appropriate treatment, mortality is reduced to 1%. *F. tularensis* is one of the most infectious pathogenic agents known, requiring inhalation or inoculation of fewer than 50 organisms to cause disease. The organism is fastidious and slow growing requiring several days to culture. Culture also poses as a risk to laboratory personnel due to the highly infectious nature of the organism and several cases of laboratory acquired infection have been documented in the literature. Serology is commonly used to diagnose tularemia however antibody response is not detectable until two or more weeks after initial infection. The highly infective nature of the organism and the seriousness of the infection it causes has made *F. tularensis* a putative biological warfare agent. Alternative more rapid and sensitive detection methods such as real-time PCR, which may remove the need for culturing the organism may prove useful for the detection of this pathogen. Emanuel *et al.* (2003) recently reported a real-time assay for the detection of *F. tularensis* which targeted the *fopA* and *tul4* genes. The assays recognised all three *F. tularensis* biogroups, type A tularensis, type B holartica, and novicida. The assays were developed on both the ABI 7900 platform and on a novel hand-held thermal cycler, the BioSeeq developed by Smiths Detection (Edgewood, Md.). This is a battery-powered compact instrument containing six independently programmable thermal cycler optics modules. This instrument was designed to be operated by first responders in the field using pre-set protocols. The limit of detection (LOD) of the assays on the ABI 7900 was approximately 25 GE with the LOD on the BioSeeq being slightly higher for each assay at 100 GE for the *tul4* assay and 150 GE for the *fopA* assay. The assays were applied to the detection of *F. tularensis* in tissues sampled over a five-day period from experimentally infected mice to determine their usefulness as diagnostic tests. The results demonstrated that the PCR assays were slightly less sensitive than culture however they did provide results in a much more timely manner, four hours compared to three days for culture. Several small and rapid, hand-held sized thermal cyclers for the use in the field by first responders have been developed. The U.S. Department of Energy's Lawrence Livermore National Laboratory developed and manufactured a suitcase-sized unit MATCI (miniature analytical thermal cycler instrument) which was used to detect and differentiate between orthopox virus species (Ibrahim *et al.*, 1998). The MATCI system unfortunately had only a single reaction chamber therefore a second generation model was developed, the ANAA (advanced nucleic acid analyser), with ten reaction modules (Belgrader *et al.*, 1998). A third generation 'hand-held' sized four reaction unit was also developed, the HANAA, (handheld advanced nucleic acid analyser) which has an extremely rapid detection time of six to ten minutes. These hand-held thermocyclers show promise as an expedient means of forward diagnosis of infection in the field.

Smallpox was a devastating disease of man that was eradicated from the world in 1977 however stocks of the virus do still exist. It is considered a potential agent of bioterrorism because of the devastating disease it causes. The threat of a deliberate release of smallpox has led efforts to replenish the supply of vaccine and develop new drugs and diagnostic tests for the rapid and early detection of infection. Two real-time PCR assays have been described for the detection of smallpox virus, a LC hybridisation assay and a 5' nuclease assay which both target the hemagglutinin gene. The assay of Espy *et al.* (2002a) demonstrated a LOD of approximately 5 to 10 GE and could differentiate between the smallpox virus, cowpox virus, monkeypox virus and vaccinia virus.

Differentiation was achieved by detection of nucleotide sequence mismatches in the probe hybridisation sequence using melting curve analysis with the different virus types having individual Tm specific for them. The real-time assay of Ibrahim *et al.* (2003) was evaluated on both the SmartCycler and LC platforms. The assay demonstrated a LOD of approximately 25 GE with variola virus genomic DNA and a LOD of approximately 12 gene copies with plasmid DNA containing the target sequence. The assay was evaluated with 322 coded samples that included variola virus-infected cells and tissues, purified variola virus DNA, and the DNA of 25 other species of orthopoxviruses, herpes simplex virus, *Rickettsia*, myxoma virus, and varicella-zoster virus. The sensitivities and specificities of the assay were 96.1% and 99.3% with the SmartCycler instrument and 99.5% and 95.7% with the LC instrument respectively. Two of the 116 samples not containing variola virus were falsely positive in the SmartCycler instrument and five samples were falsely positive in the LC. The authors concluded that these false-positives may have occurred due to cross-contamination during testing. This assay is the first to be applied to 48 different isolates of variola virus demonstrating the sensitivity of the assay for this target organism and the conserved nature of the target sequence within variola virus isolates.

The detection of bioterrorism agents by clinical microbiology laboratories has raised concerns about the safety implications of such testing. A novel system to reduce the risk to laboratory personnel was suggested by Espy *et al.* (2002). They investigated the detection of vaccinia virus, herpes simplex virus, varicella-zoster virus and *B. anthracis* DNA by LC PCR following autoclaving of the samples. They demonstrated that autoclaving swabs impregnated with suspensions of the organisms resulted in the elimination of the infectivity of the agents however DNA was still detectable by real-time PCR. The authors concluded that such a method would virtually eliminate the risk to laboratory personnel while still allowing the differentiation of smallpox virus from other dermal pathogens such as HSV and VZV by LC PCR preventing the need to refer specimens positive for agents other than smallpox to laboratories with Biosafety Level 4 facilities for testing. This pragmatic approach to screening samples may be beneficial if a deliberate release of smallpox virus or another bioterrorist agent was suspected and Biosafety Level 4 laboratories became overwhelmed with suspect specimens.

Future Developments

This chapter has illustrated some of the areas in clinical microbiology in which the introduction of real-time methods have made a dramatic impact. The application of real-time methods to aid the diagnosis and management of infectious disease is one of the most rapidly growing fields in PCR diagnostics with an ever increasing number of assays being reported in the literature every month. Many of these are 'in-house' assays making standardisation of testing methods difficult and most have no internal controls for PCR inhibition which should be a standard requirement for any assay used as a diagnostic test. Incorporation of internal controls is fundamental to good laboratory practice and is the only way to ensure a negative result in an assay is correct and not due to the sample containing the target organism rather than through inhibition of the assay. For many of the pathogens described in the literature, any one of a number of different genes in the organism have been utilised as the target sequence for the assay however there have been very few reports of comparisons between the different gene targets to determine the most appropriate target in terms of sensitivity and specificity of detection. The development of commercially produced real-time PCR reagents and kits for diagnostics in clinical microbiology may reduce some of the problems associated with standardisation. However, even as more of these kits and reagents become available they may not be used by every laboratory. This is because many laboratories may not be able to afford the additional costs associated with making the transition to using them in comparison to the relatively cheap alternative of 'in-house' assays. A number of companies have begun to produce kits for the real-time detection of a number of pathogens including; EBV, VZV, HSV, West Nile virus, dengue, *B. anthracis*, *E. histolytica*, and *Plasmodium* (Artus GmbH, Germany); Group B Streptococcus (IDI-Strep B™ Assay) and *L. pnuemophila and M.* pnuemoniae (Minerva biolabs, Germany). In the future kits for other pathogens will be developed, however most of these will target the either the most frequently occurring pathogens found in the clinical laboratory or those which are most difficult to test for currently, ensuring the commercial success of such assays. This will mean that there will always be a need for in-house assays targeting more rarely occurring pathogens or for individual laboratory-specific interests or needs.

Prior to detection by real-time assays the nucleic acid in the clinical sample needs to be extracted in order for it to be detectable in the assay. For some specimens such as CSF this may be as simple as boiling the specimen, however for more complex sample matrices such as pulmonary secretions more complex extraction procedures are necessary. Many types of clinical samples such as faeces and whole blood contain substances inhibitory to the PCR reaction and therefore the extraction method must also remove these to prevent inhibition of the PCR. Originally manual extraction methods were the only methods available to clinical laboratories however recently a number of automated or semi-automated platforms have become available for nucleic acid extraction from samples. These platforms and associated reagents will help to standardise sample extraction methods however they will require validation studies to ensure they are suitable for the wide range of sample types found in clinical microbiology such as swabs, fluids, pulmonary secretions, tissues, CSF, urine, faeces, blood, and pus. The optimum type of sample for a particular assay and a particular pathogen also needs to be established and standardised. Clinical samples may contain very low numbers of pathogenic organisms against a background flora of many other microbes. Generic nucleic acid extraction methods produce nucleic acid extracts which contain a small amount of nucleic acid from the target pathogen against a background of nucleic acid from the rest of the mixed flora in the sample. A small volume of this sample is then tested in the PCR assay (usually between 1 and 10 μl) therefore the LOD is always compromised when there are only small numbers of target organism in the original sample. Pathogen-specific sample preparation methods which concentrate the numbers of target organisms before the nucleic acid extraction or specifically concentrate the target nucleic acid during extraction will help to increase the sensitivity of detection assays.

Quality control will have an increasingly important role in the implementation of molecular diagnostic testing for the diagnosis of infectious disease. Quality control encompasses measures such as the inclusion of appropriate positive, negative, and inhibition controls in assay runs. The results of positive controls should be monitored over time to ensure the assay is performing consistently and that inter-assay reproducibility remains high. External quality control schemes will play a very crucial role to ensure high standards in molecular diagnostics in the future. The first external quality control scheme to be developed was the European Union Quality Control Concerted Action for Nucleic Acid Amplification in Diagnostic Virology. This temporary entity has been superseded by Quality Control for Molecular Diagnostics, a non-profit organisation for the standardisation and quality control of molecular diagnostics and genomic technologies (www.qcmd.org). This organisation sends out proficiency panels of simulated clinical samples containing a wide range of viral and bacterial pathogens for molecular diagnostic assays. Over 100 laboratories from more than 60 countries regularly participate in the program which is endorsed by the European Society for Clinical Virology and the European Society for Microbiology and Infectious Disease. Laboratories providing molecular diagnostic testing should participate in this scheme to ensure quality of testing.

The introduction of real-time PCR methods in clinical microbiology has improved the detection of infectious disease agents and led to improvements in patient management and care. In the future new developments in real-time molecular diagnostics will lead to further benefits to the patient consolidating the role of real-time PCR as an essential tool in the clinical microbiology laboratory.

References

Aarts, H.J., Joosten, R.G., Henkens, M.H., Stegeman, H., van Hoek, A.H. 2001. Rapid duplex PCR assay for the detection of pathogenic *Yersinia enterocolitica* strains. J. Microbiol. Methods. 47: 209-17.

Abe, A., Inoue, K., Tanaka, T., Kato, J., Kajiyama, N., Kawaguchi, R., Tanaka, S., Yoshiba, M., and Kohara, M. 1999. Quantitation of

Hepatitis B Virus genomic DNA by real-time detection PCR. J. Clin. Microbiol. 37: 2899-2903.

Apfalter, P., Barousch, W., Nehr, M., Makristathis, A., Willinger, B., Rotter, M., and Hirschl, A.M. 2003. Comparison of a new quantitative ompA-based real-time PCR TaqMan assay for detection of *Chlamydia pneumoniae* DNA in respiratory specimens with four conventional PCR assays. J. Clin. Microbiol. 41: 592-600.

Ballard, A.L., Fry, N.K., Chan, L., Surman, S.B., Lee, J.V. Harrison T.G., and Towner K.J. 2000. Detection of *Legionella pneumophila* using a real-time PCR hybridization assay. J. Clin. Microbiol. 38: 4215-4218.

Bélanger, S.D., Boissinot, M., Ménard, C., Picard, F.J., and Bergeron, M.G. 2002. Rapid detection of shiga toxin-producing bacteria in feces by multiplex PCR with molecular beacons on the Smart Cycler. J. Clin. Microbiol. 40: 1436-40.

Bélanger, S.D., Boissinot, M., Clairoux, N., Picard, F.J., and Bergeron, M.G. 2003. Rapid detection of *Clostridium difficile* in feces by real-time PCR. J. Clin. Microbiol. 41: 730-4.

Belgrader, P., Benett, W., Hadley, D., Long, G., Mariella, R. Jr., Milanovich, F., Nasarabadi, S., Nelson, W., Richards, J., and Stratton, P. 1998. Rapid pathogen detection using a microchip PCR array instrument. Clin. Chem. 44: 2191-4.

Bell, C.A., Uhl, J.R., Hadfield, T.L., David, J.C., Meyer, R.F., Smith, T.F., and Cockerill, F.R. 2002. Detection of *Bacillus anthracis* DNA by LightCycler PCR. J. Clin. Microbiol. 40: 2897-2902.

Bellin, T., Pulz, M., Matussek, A., Hempen, H.G., and Gunzer, F. 2001. Rapid detection of enterohemorrhagic *Escherichia coli* by real-time PCR with fluorescent hybridization probes. J. Clin. Microbiol. 39: 370-374.

Berger A. and Presier W. 2002. Viral genome quantification as a tool for improving patient management: the example of HIV, HBV, HCV and CMV. J. Antimicrob. Chemother. 49:713-721.

Blessmann, J., Buss, H., Nu, P.A., Dinh, B.T., Ngo, Q.T., Van, A.L., Alla, M.D., Jackson, T.F., Ravdin, J.I., and Tannich, E. 2002. Real-Time PCR for detection and differentiation of *Entamoeba histolytica* and *Entamoeba dispar* in fecal samples. J. Clin. Microbiol. 40: 4413-4417.

Brennan, R.E. and Samuel, J.E. 2003. Evaluation of *Coxiella burnetii* antibiotic susceptibilities by real-time PCR assay. J. Clin. Microbiol. 41: 1869-1874.

Bretagne, S., Durand, R., Olivi, M., Garin, J.F., Sulahian, A., Rivollet, D., Vidaud, M., and Deniau, M. 2001. Real-time PCR as a new tool for quantifying *Leishmania infantum* in liver in infected mice. Clin. Diagn. Lab. Immunol. 8: 828-31.

Broccolo, F., Locatelli, G., Sarmati, L., Piergiovanni, S., Veglia, F., Andreoni, M., Buttò, S., Ensoli, B., Lusso, P., and Malnati, M.S. 2002. Calibrated real-time PCR assay for quantitation of human herpesvirus 8 DNA in biological fluids. J. Clin. Microbiol. 40: 4652-4658.

Callahan, J.D., Wu, S.J., Dion-Schultz, A., Mangold, B.E., Peruski, L.F., Watts, D.M., Porter, K.R., Murphy, G.R., Suharyono, W., King, C.C., Hayes, C.G., and Temenak, J.J. 2001. Development and evaluation of serotype- and group-specific fluorogenic reverse transcriptase PCR (TaqMan) assays for dengue virus. J. Clin. Microbiol. 39: 4119-24.

CDC SARS Investigative Team. 2003. Outbreak of Severe Acute Respiratory Syndrome, Worldwide. M.M.W.R. 52: 226-228.

Cloud, J.L., Hymas, W.C., Turlak, A., Croft, A., Reischl, U., Daly, J.A., and Carroll, K.C. 2003. Description of a multiplex *Bordetella pertussis* and *Bordetella parapertussis* LightCycler PCR assay with inhibition control. Diagn. Microbiol. Infect. Dis. 46: 189-95.

Collot, S., Petit, B., Bordessoule, D., Alain, S., Touati, M., Denis, F., and Ranger-Rogez, S. 2002. Real-Time PCR for quantification of human herpesvirus 6 DNA from lymph nodes and saliva. J. Clin. Microbiol. 40: 2445-2451.

Corless, C.E., Guiver, M., Borrow, R., Edwards-Jones, V., Fox, A.J., and Kaczmarski, E.B. 2001. Simultaneous detection of *Neisseria meningitidis*, *Haemophilus influenzae*, and *Streptococcus pneumoniae* in suspected cases of meningitis and septicemia using real-time PCR. J. Clin. Microbiol. 39: 1553-8.

Costa, J.M., Pautas, C., Ernault, P., Foulet, F., Cordonnier, C., and Bretagne, S. 2000. Real-time PCR for diagnosis and follow-up of *Toxoplasma* reactivation after allogeneic stem cell transplantation using fluorescence resonance energy transfer hybridization probes. J. Clin. Microbiol. 38: 2929-32.

Cote, S., Abed, Y., and Boivin, G. 2003. Comparative evaluation of real-time PCR assays for detection of the human metapneumovirus. J. Clin. Microbiol. 41: 3631-5.

Cummings, K.L., and Tarleton, R.L. 2003. Rapid quantitation of *Trypanosoma cruzi* in host tissue by real-time PCR. Mol. Biochem. Parasitol. 129: 53-9.

Deguchi, T., Yoshida, T., Yokoi, S., Ito, M., Tamaki, M., Ishiko, H., and Maeda, S. 2002. Longitudinal quantitative detection by real-time PCR of *Mycoplasma genitalium* in first-pass urine of men with recurrent nongonococcal urethritis. J. Clin. Microbiol. 40: 3854-6.

Desjardin, L.E., Chen, Y., Perkins, M.D., Teixeira, L., Cave, M.D., and Eisenach, K.D. 1998. Comparison of the ABI 7700 system (TaqMan) and competitive PCR for quantification of IS6110 DNA in sputum during treatment of tuberculosis. J. Clin. Microbiol. 36: 1964-1968.

Drago, L., Lombardi, A., Vecchi, E.D., and Gismondo, M.R. 2002. Real-time PCR assay for rapid detection of *Bacillus anthracis* spores in clinical samples. J. Clin. Microbiol. 40: 4399.

Drosten, C., Göttig, S., Schilling, S., Asper, M., Panning, M., Schmitz, H., and Günther, S. 2002. Rapid detection and quantification of RNA of Ebola and Marburg Viruses, Lassa Virus, Crimean-Congo Hemorrhagic Fever Virus, Rift Valley Fever Virus, Dengue Virus, and Yellow Fever Virus by real-time reverse transcription-PCR. J. Clin. Microbiol. 40: 2323-30.

Edwards, K.J., Kaufmann, M.E., and Saunders, N. A. 2001a. Rapid and accurate identification of coagulase-negative Staphylococci by real-time PCR. J. Clin. Microbiol. 39: 3047-51.

Edwards, K.J., Metherel,l L.A., Yates, M., and Saunders, N.A. 2001b. Detection of *rpoB* mutations in *Mycobacterium tuberculosis* by biprobe analysis. J. Clin. Microbiol. 39: 3350-2.

Ellerbrook, H. Nattermann, H., Ozel, M., Beutin L., Appel, B. and Pauli G. 2002. Rapid and sensitive identification of pathogenic and apathogenic *Bacillus anthracis* by real-time PCR. FEMS Microbiol. Lett. 214: 51-59.

Emanuel, P.A., Bell, R., Dang, J.L., McClanahan, R., David, J.C., Burgess, R.J., Thompson, J., Collins, L, and Hadfield, T. 2003. Detection of *Francisella tularensis* within infected mouse tissues by using a hand-held PCR thermocycler. J. Clin. Microbiol. 41: 689-93.

Espy, M.J., Teo, R., Ross, T.K., Svien, K.A., Wold, A.D., Uhl, J.R. and Smith, T.F. 2000a. Diagnosis of Varicella-Zoster virus infections in the clinical laboratory by LightCycler PCR. J. Clin. Microbiol. 38: 3187-9.

Espy, M.J., Uhl, J.R., Mitchell, P.S., Thorvilson, J.N., Svien, K.A., Wold, A.D., and Smith, T.F. 2000b. Diagnosis of Herpes Simplex virus infections in the clinical laboratory by LightCycler PCR. J. Clin. Microbiol. 38: 795-9.

Espy, M.J., Cockerill, F.R., Meyer, R.F., Bowen, M.D., Poland, G.A., Hadfield, T.L., and Smith, T.F. 2002. Detection of Smallpox Virus DNA by LightCycler PCR. J. Clin. Microbiol. 40: 1985-1988.

Fang, H., and Hedin, G. 2003. Rapid screening and identification of methicillin-resistant *Staphylococcus aureus* from clinical samples by selective-broth and real-time PCR assay. J. Clin. Microbiol. 41: 2894-2899.

Fenollar, F., Fournier, P.E., Raoult, D., Gerolami, R., Lepidi, H., and Poyart, C. 2002. Quantitative detection of *Tropheryma whipplei* DNA by real-time PCR. J. Clin. Microbiol. 40: 1119-1120.

Fernandez, C., Boutolleau, D., Manichanh, C., Mangeney, N., Agut, H., and Gautheret-Dejean, A. 2002. Quantitation of HHV-7 genome by real-time polymerase chain reaction assay using MGB probe technology. J.Virol.Methods 106: 11-16.

Garcia de Viedma, D., del Sol Diaz Infantes, M., Lasala, F., Chaves, F., Alcala, L., Bouza, E. 2002. New real-time PCR able to detect in a single tube multiple rifampin resistance mutations and high-level isoniazid resistance mutations in *Mycobacterium tuberculosis*. J. Clin. Microbiol. 40: 988-995.

Gault, E., Michel, Y., Dehée, A., Belabani, C., Nicolas, J-C. and Garbarg-Chenon, A. 2001. Quantification of human cytomegalovirus DNA by real-time PCR. J. Clin. Microbiol. 39: 772-775.

Gautheret-Dejean, A., Manichanh, C., Thien-Ah-Koon, F., Fillet, A.M., Mangeney, N., Vidaud, M., Dhedin, N., Vernant, J.P., and Agut, H. 2002. Development of a real-time polymerase chain reaction assay for the diagnosis of human herpesvirus-6 infection and application to bone marrow transplant patients. J. Virol. Methods 100: 27-35.

Gibson, J.R., Saunders, N.A., Burke, B., and Owen, R.J. 1999. Novel method for rapid determination of clarithromycin sensitivity in *Helicobacter pylori*. J. Clin. Microbiol. 37: 3746-8.

Greiner, O., Day, P.J., Bosshard, P.P., Imeri, F., Altwegg, M., and Nadal, D. 2001. Quantitative detection of *Streptococcus pneumoniae* in nasopharyngeal secretions by real-time PCR. J. Clin. Microbiol. 39: 3129-34.

Greiner, O., Day, P.J., Altwegg, M., and Nadal, D. 2003. Quantitative detection of *Moraxella catarrhalis* in nasopharyngeal secretions by real-time PCR. J. Clin. Microbiol. 41: 1386-90.

Griscelli, F., Barrois, M., Chauvin, S., Lastere, S., Bellet, D., and Bourhis, J-H. 2001. Quantification of human cytomegalovirus DNA in bone marrow transplant recipients by real-time PCR. J. Clin. Microbiol. 39: 4362-4369.

Grisold, A.J., Leitner, E., Mühlbauer, G., Marth, E., and Kessler, H.H. 2002. Detection of methicillin-resistant *Staphylococcus aureus* and simultaneous confirmation by automated nucleic acid extraction and real-time PCR. J. Clin. Microbiol. 40: 2392-2397.

Gueudin, M., Vabret, A., Petitjean, J., Gouarin, S., Brouard, J., and Freymuth, F. 2003. Quantitation of respiratory syncytial virus RNA in nasal aspirates of children by real-time RT-PCR assay. J. Virol. Methods. 109: 39-45.

Guiver, M., Borrow, R., Marsh, J., Gray, S.J., Kaczmarski, E.B., Howells, D., Boseley, P. and Fox A.J. 2000. Evaluation of the Applied Biosystems automated Taqman polymerase chain reaction system for the detection of meningococcal DNA. FEMS Immunol. Med. Microbiol. 28: 173-179.

Hackett, S.J., Guiver M., Marsh, J., Sillis, J.A., Thomson, A.P., Kaczmarski, E.B. and Hart C.A. 2002. Meningococcal bacterial DNA load at presentation correlates with disease severity. Arch. Dis. Child. 86: 44-46.

Hardegger, D., Nadal, D., Bossart, W., Altwegg, M., and Dutly, F. 2000. Rapid detection of *Mycoplasma pneumoniae* in clinical samples by real-time PCR. J. Microbiol. Methods 41: 45-51.

Hayden, R.T., J. R. Uhl, X. Qian, M. K. Hopkins, M. C. Aubry, A. H. Limper, R. V. Lloyd, and F. R. Cockerill. 2001. Direct detection of *Legionella* species from bronchoalveolar lavage and open lung biopsy specimens: comparison of LightCycler PCR, *in situ* hybridization, direct fluorescence antigen detection, and culture. J. Clin. Microbiol. 39: 2618-2626.

He, Q., Wang, J-P., Osato, M., and Lachman, L.B. 2002. Real-time quantitative PCR for detection of *Helicobacter pylori*. J. Clin. Microbiol. 40: 3720-3728.

Heim, A., Ebnet, C., Harste, G., and Pring-Akerblom, P. 2003. Rapid and quantitative detection of human adenovirus DNA by real-time PCR. J. Med. Virol. 70: 228-239.

Hermsen, C.C., Telgt, D.S., Linders, E.H., van de Locht, L.A., Eling, W.M., Mensink, E.J., and Sauerwein, R.W. 2001. Detection of *Plasmodium falciparum* malaria parasites *in vivo* by real-time quantitative PCR. Mol. Biochem. Parasitol. 118: 247-51.

Higgins, J.A., Ezzell, J., Hinnebusch, B.J., Shipley, M., Henchal, E.A., and Ibrahim, M.S. 1998. 5' nuclease PCR assay to detect *Yersinia pestis*. J. Clin. Microbiol. 36: 2284-8.

Higgins, J.A., Fayer, R., Trout, J.M., Xiao, L., Lal, A.A., Kerby, S., and Jenkins, M.C. 2001. Real-time PCR for the detection of *Cryptosporidium parvum*. J. Microbiol. Methods. 47: 323-37.

Hu, A., Colella, M., Tam, J.S., Rappaport, R., and Cheng, S.M. 2003. Simultaneous detection, subgrouping, and quantitation of Respiratory Syncytial Virus A and B by Real-Time PCR. J. Clin. Microbiol. 41: 149-154.

Ibrahim, M.S., Lofts, R.S., Jahrling, P.B., Hencha,l E.A., Weedn, V.W., Northrup, M.A., and Belgrader, P. 1998. Real-time microchip PCR for detecting single-base differences in viral and human DNA. Anal. Chem. 70: 2013-7.

Jauregui, L.H., Higgins, J., Zarlenga, D., Dubey, J.P., and Lunney, J.K. 2001. Development of a real-time PCR assay for detection of *Toxoplasma gondii* in pig and mouse tissues. J. Clin. Microbiol. 39: 2065-71.

Kageyama, T., Kojima, S., Shinohara, M., Uchida, K., Fukushi, S., Hoshino, F.B., Takeda, N., and Katayama, K. 2003. Broadly reactive and highly sensitive assay for Norwalk-Like viruses based on real-time quantitative reverse transcription-PCR. J. Clin. Microbiol. 41: 1548-1557.

Kearns, A.M., Turner, A.J., Taylor, C.E., George, P.W., Freeman, R., and Gennery, A.R. 2001a. LightCycler-based quantitative PCR for rapid detection of human herpesvirus 6 DNA in clinical material. J. Clin. Microbiol. 39: 3020-1.

Kearns, A.M., Guiver, M., James, V., and King, J. 2001b. Development of a real-time quantitative PCR for the detection of human cytomegalovirus. J. Virol. Methods 95: 121-131.

Kearns, A.M., Draper, B., Wipat, W., Turner, A.J., Wheeler, J., Freeman, R., Harwood, J., Gould, F.K., and Dark, J.H. 2001c. LightCycler-based quantitative PCR for detection of cytomegalovirus in blood, urine, and respiratory samples. J. Clin. Microbiol. 39: 2364-5.

Kearns, A.M., Graham, C., Burdess, D., Heatherington, J., and Freeman, R. 2002a. Rapid real-time PCR for determination of penicillin susceptibility in pneumococcal meningitis, including culture-negative cases. J. Clin. Microbiol. 40: 682-684.

Kearns, A.M., Turner, A.J., Eltringham, G.J., and Freeman, R. 2002b. Rapid detection and quantification of CMV DNA in urine using LightCycler-based real-time PCR. J. Clin. Virol. 24: 131-134.

Kessler, H.H., Mühlbauer, G., Rinner, B., Stelzl, E., Berger, A., Dörr, H-W., Santner, B., Marth, E., and Rabenau, H. 2000. Detection of Herpes Simplex Virus DNA by real-time PCR. J. Clin. Microbiol. 38: 2638-2642.

Killgore, G.E., Holloway, B., and Tenover, F.C. 2000. A 5' nuclease PCR (TaqMan) high-throughput assay for detection of the *mecA* gene in staphylococci. J. Clin. Microbiol. 38: 2516-9.

Kimura, H., Morita, M., Yabuta, Y., Kuzushima, K., Kato, K., Kojima, S., Matsuyama, T. and Morishima, T. 1999. Quantitative analysis of Epstein-Barr virus load by using a real-time PCR assay. J. Clin. Microbiol. 37: 132-136.

Kleiber, J., Walter, T., Haberhausen, G., Tsang, S., Babiel, R., and Rosenstraus, M. 2000. Performance characteristics of a quantitative, homogeneous TaqMan RT-PCR test for HCV RNA. J. Mol. Diagn. 2: 158-66.

Kösters, K., Riffelmann, M., and Wirsing von Konig, C.H. 2001. Evaluation of a real-time PCR assay for detection of *Bordetella pertussis* and *B. parapertussis* in clinical samples. J. Med. Microbiol. 50: 436-40.

Kösters, K., Reischl, U., Schmetz, J., Riffelmann, M., and Wirsing von König, C.H. 2002. Real-Time LightCycler PCR for detection and discrimination of *Bordetella pertussis* and *Bordetella parapertussis*. J. Clin. Microbiol. 40: 1719-1722.

Kuoppa, Y., Boman, J., Scott, L., Kumlin, U., Eriksson, I., and Allard, A. 2002. Quantitative detection of respiratory *Chlamydia pneumoniae* infection by Real-Time PCR. J. Clin. Microbiol. 40: 2273-2274.

Kupferschmidt, O., Kruger, D., Held, T.K., Ellerbrok, H., Siegert, W., and Janitschke, K. 2001. Quantitative detection of *Toxoplasma gondii* DNA in human body fluids by TaqMan polymerase chain reaction. Clin. Microbiol. Infect. 7: 120-4.

Kraus, G., Cleary, T., Miller, N., Seivright, R., Young, A.K., Spruill, G., and Hnatyszyn, H.J. 2001. Rapid and specific detection of the *Mycobacterium tuberculosis* complex using fluorogenic probes and real-time PCR. Mol. Cell. Probes. 15: 375-83.

Lai, K.K., Cook, L., Wendt, S., Corey, L., and Jerome, K.R. 2003. Evaluation of real-time PCR versus PCR with liquid-phase hybridization for detection of Enterovirus RNA in cerebrospinal fluid. J. Clin. Microbiol. 41: 3133-3141.

Lanciotti, R.S., Kerst, A.J., Nasci, R.S., Godsey, M.S., Mitchell, C.J., Savage, H.M., Komar, N., Panella, N.A., Allen, B.C., Volpe, K.E., Davis, B.S., and Roehrig, J.T. 2000. Rapid detection of West Nile virus from human clinical specimens, field-collected mosquitoes, and avian samples by a TaqMan reverse transcriptase-PCR assay. J. Clin. Microbiol. 38: 4066-71.

Lapierre, P., Huletsky, A., Fortin, V., Picard, F.J., Roy, P.H., Ouellette, M., and Bergeron, M.J. 2003. Real-time PCR assay for detection of fluoroquinolone resistance associated with *grlA* mutations in *Staphylococcus aureus*. J. Clin. Microbiol. 41: 3246-3251.

Lee, M.A., Brightwell, G., Leslie, D., Bird, H., and Hamilton, A. 1999. Fluorescent detection techniques for real-time multiplex strand specific detection of *Bacillus anthracis* using rapid PCR. J. Appl. Microbiol. 87: 218-23.

Lee, M-A., Tan, C-H., Aw, L-T., Tang, C-S., Singh, M., Lee, S-H., Chia, H-P., and Yap, E.P.H. 2002. Real-time fluorescence-based PCR for detection of malaria parasites. J. Clin. Microbiol. 40: 4343-4345.

Leruez-Ville, M., Ouachée, M., Delarue, R., Sauget, A-S., Blanche, S., Buzyn, A., and Rouzioux, C. 2003. Monitoring Cytomegalovirus infection in adult and pediatric bone marrow transplant recipients by a real-time PCR assay performed with blood plasma. J. Clin. Microbiol. 41: 2040-2046.

Li, H., Dummer, J.S., Estes, W.R., Meng, S., Wright, P.F., and Tang, Y.W. 2003. Measurement of human Cytomegalovirus loads by quantitative real-time PCR for monitoring clinical intervention in transplant recipients. J. Clin. Microbiol. 41: 187-191.

Lindler, L.E., Fan, W., and Jahan, N. 2001. Detection of ciprofloxacin-resistant *Yersinia pestis* by fluorogenic PCR using the LightCycler. J. Clin. Microbiol. 39: 3649-55.

Limor, J.R., Lal, A.A, and Xiao, L. 2002. Detection and differentiation of Cryptosporidium parasites that are pathogenic for humans by real-time PCR. J. Clin. Microbiol. 40: 2335-2338.

Lin, M.H., Chen, T.C., Kuo, T.T., Tseng, C.C., and Tseng, C.P. 2000. Real-time PCR for quantitative detection of *Toxoplasma gondii*. J. Clin. Microbiol. 38: 4121-4125.

Liveris, D., Wang, G., Girao, G., Byrne, D. W., Nowakowski, J., McKenna, D., Nadelman, R., Wormser, G. P., and Schwartz, I. 2002. Quantitative detection of *Borrelia burgdorferi* in 2-millimeter skin samples of erythema migrans lesions: correlation of results with clinical and laboratory findings. J. Clin. Microbiol. 40: 1249-1253.

Locatelli, G., Santoro, F., Veglia, F., Gobbi, A., Lusso, P., and Malnati, M.S. 2000. Real-time quantitative PCR for human herpesvirus 6 DNA. J. Clin. Microbiol. 38: 4042-8.

Logan, J. M. J., Edwards, K. J., Saunders, N. A., and Stanley, J. 2001. Rapid identification of *Campylobacter* spp. by melting peak analysis of biprobes in real-time PCR. J. Clin. Microbiol. 39: 2227-32.

Machida, U., Kami, M., Fukui, T., Kazuyama, Y., Kinoshita, M., Tanaka, Y., Kanda, Y., Ogawa, S., Honda, H., Chiba, S., Mitani, K., Muto, Y., Osumi, K., Kimura, S., and Hirai, H. 2000. Real-time automated PCR for early diagnosis and monitoring of Cytomegalovirus infection after bone marrow transplantation. J. Clin. Microbiol. 38: 2536-2542.

Mackay, I.M., Arden, K.E., and Nitsche, A. 2002. Real-time PCR in virology. Nucleic Acid Res. 30: 1292-305.

Mackay, I.M., Jacob, K.C., Woolhouse, D., Waller, K., Syrmis, M.W., Whiley, D.M., Siebert, D.J., Nissen, and M., Sloots, T.P. 2003. Molecular assays for detection of human metapneumovirus. J. Clin. Microbiol. 41: 100-5.

Makinen, J., Mertsola, J., Viljanen, M.K,, Arvilommi, H., and He, Q. 2002. Rapid typing of *Bordetella pertussis* pertussis toxin gene variants by LightCycler real-time PCR and fluorescence resonance energy transfer hybridization probe melting curve analysis. J. Clin. Microbiol. 40: 2213-2216.

Makino, S.I., Cheun, H.I., Watarai, M., Uchida, I., and Takeshi, K. 2001. Detection of anthrax spores from the air by real-time PCR. Lett. Appl. Microbiol. 33: 237-40.

Makino, S., and Cheun, H.I. 2003. Application of the real-time PCR for the detection of airborne microbial pathogens in reference to the anthrax spores. J. Microbiol. Methods. 53: 141-7.

Martell, M., Gómez, J., Esteban, J.L., Sauleda, S., Quer, J., Cabot, B., Esteban, R., and Guardia, J. 1999. High-throughput real-time reverse transcription-PCR quantitation of Hepatitis C virus RNA. J. Clin. Microbiol. 37: 327-332.

Matsumura, M., Y.Hikiba, K. Ogura, G. Togo, I. Tsukuda, K. Ushikawa, Y. Shiratori, and M. Omata. 2001. Rapid detection of mutations in the 23S rRNA gene of *Helicobacter pylori* that confers resistance to clarithromycin treatment to the bacterium. J. Clin. Microbiol. 39: 691-5.

McAvin, J.C., Reilly, P.A., Roudabush, R.M., Barnes, W.J., Salmen, A., Jackson, G.W., Beninga, K.K., Astorga, A., McCleskey, F.K., Huff, W.B., Niemeyer, D., and Lohman, K.L. 2001. Sensitive and specific method for rapid identification of *Streptococcus pneumoniae* using real-time fluorescence PCR. J. Clin. Microbiol. 39: 3446-3451.

Mengelle, C., Sandres-Saune, K., Pasquier, C., Rostaing, L., Mansuy, J.M., Marty, M., Da Silva, I., Attal, M., Massip, P. and Izopet, J. 2003. Automated extraction and quantification of human cytomegalovirus DNA in whole blood by real-time PCR assay. J. Clin. Microbiol. 41: 3840-3845.

Menotti, J., Cassinat, B., Sarfati, C., Liguory, O., Derouin, F., and Molina, J.M. 2003. Development of a real-time PCR assay for quantitative detection of *Encephalitozoon intestinalis* DNA. J. Clin. Microbiol. 41: 1410-3.

Miller, N., Cleary, T., Kraus, G., Young, A.K., Spruill, G. and Hnatyszyn H.J. 2002. Rapid and specific detection of *Mycobacterium tuberculosis* from acid-fast bacillus smear-positive respiratory specimens and BacT/ALERT MP culture bottles by using fluorogenic probes and real-time PCR. J. Clin. Microbiol. 40: 4143-4147.

Molling, P., Jacobsson, S., Backman, A., and Olcen, P. 2002. Direct and rapid identification and genogrouping of meningococci and *porA* amplification by LightCycler PCR. J. Clin. Microbiol. 40: 4531-4535.

Mommert, S., Gutzmer, R., Kapp, A., and Werfel, T. 2001. Sensitive detection of *Borrelia burgdorferi* sensu lato DNA and differentiation of *Borrelia* species by LightCycler PCR. J. Clin. Microbiol. 39: 2663-7.

Monpoeho, S., Coste-Burel, M., Costa-Mattioli, M., Besse, B.,

Chomel, J.J., Billaudel, S., and Ferre V. 2002. Application of a real-time polymerase chain reaction with internal positive control for detection and quantification of enterovirus in cerebrospinal fluid. Eur. J. Clin. Microbiol. Infect. Dis. 21: 532-6.

Morrison, T.B., Ma, Y., Weis, J.H., and Weis, J.J. 1999. Rapid and sensitive quantification of *Borrelia burgdorferi*-infected mouse tissues by continuous fluorescent monitoring of PCR. J. Clin. Microbiol. 37: 987-92.

Mothershed, E.A. Cassiday, P.K., Pierson, K., Mayer, L.W., and Popovic, T. 2002. Development of a real-time fluorescence PCR assay for rapid detection of the diphtheria toxin gene. J. Clin. Microbiol. 40: 4713-4719.

Nicolas, L., Prina, E., Lang, T., and Milon, G. 2002. Real-time PCR for detection and quantitation of leishmania in mouse tissues. J. Clin. Microbiol. 40: 1666-9.

Nielsen, E.M., and Andersen, M.T. 2003. Detection and characterization of verocytotoxin-producing *Escherichia coli* by automated 5' nuclease PCR assay. J. Clin. Microbiol. 41: 2884-2893.

Niesters, H.G., van Esser, J., Fries, E., Wolthers, K.C., Cornelissen, J., and Osterhaus, A.D. 2000. Development of a real-time quantitative assay for detection of Epstein-Barr Virus. J. Clin. Microbiol. 38: 712-5.

Niesters, H.G. 2001. Quantitation of viral load using real-time amplification techniques. Methods. 25: 419-29.

Niesters, H.G. 2002. Clinical virology in real time. J. Clin. Virol. 25: S3-12.

Nijhuis, M., van Maarseveen, N., Schuurman, R., Verkuijlen, S., de Vos, M., Hendriksen, K., and van Loon, A.M. 2002. Rapid and sensitive routine detection of all members of the genus Enterovirus in different clinical specimens by real-time PCR. J. Clin. Microbiol. 40: 3666-70.

Nitsche, A., Steuer, N., Schmidt, C.A., Landt, O., Ellerbrok, H., Pauli, G., and Siegert, W. 2000. Detection of human Cytomegalovirus DNA by real-time quantitative PCR. J. Clin. Microbiol. 38: 2734-2737.

Nogva, H.K., Bergh, A., Holck, A., and Rudi, K. 2000. Application of the 5'-nuclease PCR assay in evaluation and development of methods for quantitative detection of *Campylobacter jejuni*. Appl. Environ. Microbiol. 66: 4029-36.

Oberst, R.D., Hays, M.P., Bohra, L.K., Phebus, R.K., Yamashiro, C.T., Paszko-Kolva, C., Flood, S.J., Sargeant, J.M., and Gillespie, J.R. 1998. PCR-based DNA amplification and presumptive detection of *Escherichia coli* O157: H7 with an internal fluorogenic probe and the 5' nuclease (TaqMan) assay. Appl. Environ. Microbiol. 64: 3389-96.

Oggioni, M.R., Meacci, F., Carattoli, A., Ciervo, A., Orru, G., Cassone, A., and Pozzi, G. 2002. Protocol for real-time PCR identification of anthrax spores from nasal swabs after broth enrichment. J. Clin. Microbiol. 40: 3956-63.

Oleastro, M., Ménard, A., Santos, A., Lamouliatte, H., Monteiro, L., Barthélémy, P., and Mégraud, F. 2003. Real-Time PCR assay for rapid and accurate detection of point mutations conferring resistance to clarithromycin in *Helicobacter pylori*. J. Clin. Microbiol. 41: 397-402.

Pahl, A., Kühlbrandt, U., Brune, K., Röllinghoff, M., and Gessner, A. 1999. Quantitative detection of *Borrelia burgdorferi* by real-time PCR. J. Clin. Microbiol. 37: 1958-1963.

Palladino, S., Kay, I.D., Costa, A.M., Lambert, E.J., and Flexman, J.P. 2003a. Real-time PCR for the rapid detection of *van*A and *van*B genes. Diagn. Microbiol. Infect. Dis. 45: 81-4.

Palladino, S., Kay, I.D., Flexman, J.P., Boehm, I., Costa, A.M.G., Lambert, E.J. and Christiansen, K. J. 2003b. Rapid detection of *van*A and *van*B genes directly from clinical specimens and enrichment broths by real-time multiplex PCR Assay. J. Clin. Microbiol. 41: 2483-2486.

Palomares, C., Torres, M.J., Torres, A., Aznar, J., and Palomares, J.C. 2003. Rapid detection and identification of *Staphylococcus aureus* from blood culture specimens using real-time fluorescence PCR. Diagn. Microbiol. Infect. Dis. 45(3): 183-9.

Pas, S.D., Fries, E., De Man, R.A., Osterhaus, A.D., and Niesters,

H.G. 2000. Development of a quantitative real-time detection assay for Hepatitis B Virus DNA and comparison with two commercial assays. J. Clin. Microbiol. 38: 2897-2901.

Patel, S., Zuckerman, M., and Smith, M. 2003. Real-time quantitative PCR of Epstein-Barr virus BZLF1 DNA using the LightCycler. J. Virol. Methods. 109: 227-33.

Piesman, J., Schneider, B.S., and Zeidner, N.S. 2001. Use of quantitative PCR to measure density of *Borrelia burgdorferi* in the midgut and salivary glands of feeding tick vectors. J. Clin. Microbiol. 39: 4145-8.

Pietila, J., He, Q., Oksi, J., and Viljanen, M.K. 2000. Rapid differentiation of *Borrelia garinii* from *Borrelia afzelii* and *Borrelia burgdorferi* Sensu Stricto by LightCycler fluorescence melting curve analysis of a PCR product of the *recA* Gene. J. Clin. Microbiol. 38: 2756-2759.

Qi, Y., Patra, G., Liang, X., Williams, L.E., Rose, S., Redkar, R.J. and Vito G. DelVecchio. 2001. Utilization of the *rpo*B gene as a specific chromosomal marker for real-time PCR detection of *Bacillus anthracis*. Appl. Envir. Microbiol. 67: 3720-3727.

Rabenau, H.F., Clarici, A.M., Muhlbauer, G., Berger, A., Vince, A., Muller, S., Daghofer, E., Santner, B.I., Marth, E., and Kessler, H.H. 2002. Rapid detection of enterovirus infection by automated RNA extraction and real-time fluorescence PCR. J. Clin. Virol. 25: 155-64.

Rantakokko-Jalava, K., and Jalava, J. 2001. Development of Conventional and Real-Time PCR Assays for detection of *Legionella* DNA in respiratory specimens. J. Clin. Microbiol. 39: 2904-2910.

Rauter, C., Oehme, R., Diterich, I., Engele, M., and Hartung, T. 2002. Distribution of clinically relevant *Borrelia* genospecies in ticks assessed by a novel, single-run, real-time PCR. J. Clin. Microbiol. 40: 36-43.

Reischl, U., Linde, H-J., Metz, M., Leppmeier, B., and Lehn N. 2000. Rapid identification of methicillin-resistant *Staphylococcus aureus* and simultaneous species confirmation using real-time fluorescence PCR. J. Clin. Microbiol. 38: 2429-2433.

Reischl, U., Lehn, N., Sanden, G.N., and Loeffelholz, M.J. 2001. Real-time PCR assay targeting IS481 of *Bordetella pertussis* and molecular basis for detecting *Bordetella holmesii*. J. Clin. Microbiol. 39: 1963-66.

Reischl, U., Youssef, M.T., Kilwinski, J., Lehn, N., Zhang, W.L., Karch, H., and Strockbine, N.A. 2002a. Real-time fluorescence PCR assays for detection and characterization of Shiga toxin, Intimin, and Enterohemolysin genes from shiga toxin-producing *Escherichia coli*. J. Clin. Microbiol. 40: 2555-2565.

Reischl, U., Linde, H.J., Lehn, N., Landt, O., Barratt, K., and Wellinghausen, N. 2002b. Direct detection and differentiation of *Legionella* spp. and *Legionella pneumophila* in clinical specimens by dual-color real-time PCR and melting curve analysis. J. Clin. Microbiol. 40: 3814-3817.

Reischl, U., Bretagne, S., Kruger, D., Ernault, P., and Costa, J.M. 2003a. Comparison of two DNA targets for the diagnosis of Toxoplasmosis by real-time PCR using fluorescence resonance energy transfer hybridization probes. BMC Infect Dis. 3: 7.

Reischl, U., Lehn, N., Simnacher, U., Marre, R., and Essig, A. 2003b. Rapid and standardized detection of *Chlamydia pneumoniae* using LightCycler real-time fluorescence PCR. Eur. J. Clin. Microbiol. Infect Dis. 22: 54-7.

Ryncarz, A.J., Goddard, J., Wald, A., Huang, M.L., Roizman, B., Corey, L. 1999. Development of a high-throughput quantitative assay for detecting Herpes Simplex Virus DNA in clinical samples. J. Clin. Microbiol. 37: 1941-1947.

Safronetz, D., Humar, A., and Tipples, G.A. 2003. Differentiation and quantitation of human herpesviruses 6A, 6B and 7 by real-time PCR. J. Virol. Methods 112: 99-105.

Sails, A.D., Fox, A.J., Bolton, F. J., Wareing, D.R.A. and Greenway, D.L.A. 2003. A real-time PCR Assay for the detection of *Campylobacter jejuni* in foods after enrichment culture. Appl. Environ. Microbiol. 69: 1383-1390.

Sanchez, J.L. and Storch, G.A. 2002. Multiplex, quantitative, real-time PCR assay for Cytomegalovirus and human DNA. J. Clin. Microbiol. 40: 2381-2386.

Schulz, A., Mellenthin, K., Schönian, G., Fleischer, B., and Drosten, C. 2003. Detection, differentiation, and quantitation of pathogenic *Leishmania* organisms by a fluorescence resonance energy transfer-based real-time PCR assay. J. Clin. Microbiol. 41: 1529-1535.

Sharma, V.K., Dean-Nystrom, E.A., and Casey, T.A. 1999. Semi-automated fluorogenic PCR assays (TaqMan) for rapid detection of *Escherichia coli* O157: H7 and other shiga toxigenic *E. coli*. Mol. Cell. Probes. 13: 291-302.

Sharma, V.K., and Dean-Nystrom, E.A. 2003. Detection of enterohemorrhagic *Escherichia coli* O157: H7 by using a multiplex real-time PCR assay for genes encoding intimin and Shiga toxins. Vet Microbiol. 93: 247-60.

Shu, P.Y., Chang, S.F., Kuo, Y.C., Yueh, Y.Y., Chien, L.J., Sue, C.L., Lin, T.H., and Huang, J.H. 2003. Development of group- and serotype-specific one-Step SYBR Green I based real-time reverse transcription PCR assay for Dengue Virus. J. Clin. Microbiol. 41: 2408-2416.

Sloan, L.M., Hopkins, M.K., Mitchell, P.S., Vetter, E.A. Rosenblatt, J.E., Harmsen, W.S., Cockerill, F.R. and Patel, R. 2002. Multiplex LightCycler PCR assay for detection and differentiation of *Bordetella pertussis* and *Bordetella parapertussis* in nasopharyngeal specimens. J. Clin. Microbiol. 40: 96-100.

Sofi Ibrahim, M., Kulesh, D.A., Saleh, S.S., Damon, I.K., Esposito, J.J., Schmaljohn, A.L., and Jahrling, P.B. 2003. Real-time PCR assay to detect smallpox virus. J. Clin. Microbiol. 41: 3835-9.

Stamey, F.R., Patel, M.M., Holloway, B.P., and Pellett, P.E. 2001. Quantitative, fluorogenic probe PCR assay for detection of human herpesvirus 8 DNA in clinical specimens. J. Clin. Microbiol. 39: 3537-40.

Stevens, S.J., Verkuijlen, S.A., Brule, A.J., Middeldorp, J.M. 2002. Comparison of quantitative competitive PCR with LightCycler-based PCR for measuring Epstein-Barr Virus DNA load in clinical specimens. J. Clin. Microbiol. 40: 3986-3992.

Stocher, M., and Berg, J. 2002. Normalized quantification of human cytomegalovirus DNA by competitive real-time PCR on the

LightCycler instrument. J. Clin. Microbiol. 40: 4547-53.

Tan, T.Y., Corden, S., Barnes, R., and Cookson, B. 2001. Rapid identification of methicillin-resistant *Staphylococcus aureus* from positive blood cultures by real-time fluorescence PCR. J. Clin. Microbiol. 39: 4529-4531.

Tanriverdi, S., Tanyeli, A., Baslamisli, F., Köksal, F., Kilinç, Y., Feng, X. Batzer, G., Tzipori, S., and Widmer, G. 2002. Detection and genotyping of oocysts of *Cryptosporidium parvum* by real-time PCR and melting curve analysis. J. Clin. Microbiol. 40: 3237-3244.

Tomaso, H., Reisinger, E.C., Al Dahouk, S., Frangoulidis, D., Rakin, A., Landt, O., and Neubauer, H. 2003. Rapid detection of *Yersinia pestis* with multiplex real-time PCR assays using fluorescent hybridisation probes. FEMS Immunol. Med. Microbiol. 38: 117-26.

Tondella, M.L.C., Talkington, D.F., Holloway, B.P., Dowell, S.F., Cowley, K., Soriano-Gabarro, M., Elkind, M.S., and Fields, B.S. 2002. Development and evaluation of real-time PCR-based fluorescence assays for detection of *Chlamydia pneumoniae*. J. Clin. Microbiol. 40: 575-583.

Torres, M.J., Criado, A., Palomares, J.C., and Aznar, J. 2000. Use of real-time PCR and fluorimetry for rapid detection of rifampin and isoniazid resistance-associated mutations in *Mycobacterium tuberculosis*. J. Clin. Microbiol. 38: 3194-3199.

Uhl, J.R., Adamson, S.C., Vetter, E.A., Schleck, C.D., Harmsen, W.S., Iverson, L.K., Santrach, P.J., Henry, N.K., and Cockerill, F.R. 2003. Comparison of LightCycler PCR, rapid antigen immunoassay, and culture for detection of group A streptococci from throat swabs. J. Clin. Microbiol. 41: 242-9.

van Doornum, G.J., Guldemeester, J., Osterhaus, A.D., and Niesters, H.G. 2003. Diagnosing herpesvirus infections by real-time amplification and rapid culture. J. Clin. Microbiol. 41: 576-80.

van Elden, L.J.R., Nijhuis, M., Schipper, P., Schuurman, R., and van Loon, A.M. 2001. Simultaneous detection of Influenza Viruses A and B using real-time quantitative PCR. J. Clin. Microbiol. 39: 196-200.

Varma, M., J.D. Hester, F.W. Schaefer, M.W., Ware, and H.D. Lindquist. 2003. Detection of *Cyclospora cayetanensis* using a quantitative real-time PCR assay. J. Microbiol. Methods. 53: 27-36.

Verstrepen, W. A., Kuhn, S., Kockx, M.M., Van De Vyvere, M.E., and Mertens, A.H. 2001. Rapid detection of Enterovirus RNA in cerebrospinal fluid specimens with a novel single-tube real-time reverse transcription-PCR assay. J. Clin. Microbiol. 39: 4093-4096.

Walker, R.A., Saunders, N., Lawson, A.J., Lindsay, E.A., Dassama, M., Ward, L.R., Woodward, M.J., Davies, R.H., Liebana, E., and Threlfall, E.J. 2001. Use of a LightCycler *gyrA* mutation assay for rapid identification of mutations conferring decreased susceptibility to ciprofloxacin in multiresistant *Salmonella enterica* serotype *Typhimurium* DT104 isolates. J. Clin. Microbiol. 39: 1443-8.

Wang, G., Liveris, D., Brei, B., Wu, H., Falco, R. C., Fish, D. and Schwartz, I. 2003. Real-Time PCR for simultaneous detection and quantification of *Borrelia burgdorferi* in field-collected *Ixodes scapularis* ticks from the Northeastern United States. *Appl. Environ. Microbiol.* 69: 4561-4565

Weidmann, M., Meyer-König, U., and Hufert, F.T. 2003. Rapid detection of Herpes Simplex virus and Varicella-Zoster virus infections by real-time PCR. J. Clin. Microbiol. 41: 1565-1568.

Wellinghausen, N., Frost, C., and Marre, R. 2001. Detection of *Legionellae* in hospital water samples by quantitative real-time LightCycler PCR. Appl. Envir. Microbiol. 67: 3985-3993.

Welti, M., Jaton, K., Altwegg, M., Sahli, R., Wenger, A., and Bille, J. 2003. Development of a multiplex real-time quantitative PCR assay to detect *Chlamydia pneumoniae*, *Legionella pneumophila* and *Mycoplasma pneumoniae* in respiratory tract secretions. Diagn. Microbiol. Infect. Dis. 45: 85-95.

Wilson, D.L., Abner, S.R., Newman, T.C., Mansfield, L.S., and Linz, J.E. 2000. Identification of ciprofloxacin-resistant *Campylobacter jejuni* by use of a fluorogenic PCR assay. J. Clin. Microbiol. 38: 3971-3978.

Wilson, D.A., Yen-Lieberman, B., Reischl, U., Gordon, S.M. and Procop, G.W. 2003. Detection of *Legionella pneumophila* by real-time PCR for the *mip* Gene. J. Clin. Microbiol. 41: 3327-3330.

Witney, A.A., Doolan, D.L., Anthony, R.M., Weiss, W.R., Hoffman, S.L., and Carucci, D.J. 2001. Determining liver stage parasite burden by real time quantitative PCR as a method for evaluating pre-erythrocytic malaria vaccine efficacy. Mol. Biochem. Parasitol. 118: 233-45.

Woodford, N., Tysall, L., Auckland, C., Stockdale, M.W., Lawson, A.J., Walker, R.A., and Livermore, D.M. 2002. Detection of oxazolidinone-resistant *Enterococcus faecalis* and *Enterococcus faecium* strains by real-time PCR and PCR-restriction fragment length polymorphism analysis. J. Clin. Microbiol. 40: 4298-300.

Wolk, D.M., Schneider, S.K., Wengenack, N.L., Sloan, L.M., and Rosenblatt, J.E. 2002. Real-time PCR method for detection of *Encephalitozoon intestinalis* from stool specimens. J. Clin. Microbiol. 40: 3922-8.

Wortmann, G., Sweeney, C. Houng, H.S., Aronson, N., Stiteler, J., Jackson, J., and Ockenhouse, C. 2001. Rapid diagnosis of leishmaniasis by fluorogenic polymerase chain reaction. Am. J. Trop. Med. Hyg. 65: 583-7.

Yang, J.H., Lai, J.P., Douglas, S.D., Metzger, D., Zhu, X.H., and Ho, W.Z. 2002. Real-time RT-PCR for quantitation of hepatitis C virus RNA. J. Virol. Methods. 102: 119-28.

Yoshida, T., Deguchi, T., Ito, M., Maeda, S., Tamaki, M., and Ishiko, H. 2002. Quantitative detection of *Mycoplasma genitalium* from first-pass urine of men with urethritis and asymptomatic men by real-time PCR. J. Clin. Microbiol. 40: 1451-1455.

Zeaiter, Z., Fournier, P.E., Greub, G. and Raoult, D. 2003. Diagnosis of Bartonella endocarditis by a real-time nested PCR assay using serum. J. Clin. Microbiol. 41: 919-925.

Zeidner, N.S., Schneider, B.S., Dolan, M.C., and Piesman, J. 2001. An analysis of spirochete load, strain, and pathology in a model of tick-transmitted Lyme borreliosis. Vector Borne Zoonotic Dis. 1: 35-44.

12

Application of Real-Time PCR to the Diagnosis of Invasive Fungal Infection

N. Isik and N. A. Saunders

Abstract

The management of invasive fungal infections has been hampered by the inability to make a diagnosis at an early stage of the disease. Molecular diagnosis by PCR appears very promising since fungal DNA can be detected in the blood of infected patients earlier than when using conventional methods. Recently, interest in the diagnosis of invasive fungal infections by real-time PCR has increased. Real-time methods also have quantitative properties and are useful both for initial diagnosis and to assess the response to treatment. Many recent studies have combined serological tests with measurement of fungal DNA by using real-time PCR. Real-time PCR helps early diagnosis and arrangement of treatment protocols for patients with high risk of fungal infection. Here real-time PCR methods for diagnosis of invasive fungal infections are described and discussed.

Introduction

Rapid tests with high specificity and sensitivity are needed for early diagnosis of invasive fungal infections, which have non-specific clinical signs. The major groups of organisms involved are the *Aspergillus* and *Candida* spp. Blood culture and serology are the conventional methods used for diagnosis of these fungi but they have limited sensitivity, specificity and do not give a rapid indication of current infection. Consequently, ELISAs for detection of antigens (*e.g.* galactomannan and glucan) and nucleic acid detection techniques have come into use for the differential diagnosis and follow-up of fungal infections.

Tests for fungal infection are especially important for transplant recipients and patients with hematological malignancies, who have high mortality and morbidity ratios. New advances in cancer treatment and intense supportive treatment provide longer life for cancer patients but morbidity and mortality ratios increase due to invasive fungal infections. Fungal infection is now the cause of death in about half of the patients with acute leukemia. The most important factor for successful antifungal treatment in cancer patients, is to make a positive diagnosis but this may be difficult. Atypical clinical findings, difficulty in taking samples and insufficient diagnostic methods are basic problems. Although certain clinical signs and symptoms found in appropriate patient groups might indicate fungal infection the correlations are not strong. Different fungal and bacterial infections and even non-infectious conditions may cause the same clinical appearance. For many years culture has been the gold standard for diagnosis of fungal infection but taking samples for culture is problematic. Deep tissue biopsies for culture from patients with thrombocytopenia or pulmonary infiltration are particularly difficult to take. Failure to make an accurate diagnosis may prevent the administration of optimal treatment (Anaissie, 1992; Einsele *et al.*, 1997; Denning, 1998; Van Burik *et al.*, 1998).

Although the use of NASBA for detection of 18S rRNA is promising, most of the nucleic acid-based methods use PCR for detection of fungal DNA and the results have been encouraging (Kami *et al.*, 2001; Loeffler *et al.*, 2001). Often the primers are directed to conserved genes so that a wide range of fungal pathogens may be detected in a single PCR reaction (Skladny *et al.*, 1999). Several studies have shown that PCR has greater sensitivity and specificity than measurement of the plasma $(1\rightarrow3)$-β-D-glucan (BDG), latex agglutination tests, ELISA or culture. The main difficulties have been the non-quantitative nature of the PCR results and contamination in reactions that use primers targeting conserved genes (Loeffler *et al.*, 1999). In comparison with the conventional PCRs the antigen detection tests are quantitative within a narrow concentration range (Becker *et al.*, 2000) and may therefore be superior for monitoring invasive disease.

The PCR based-methods are now being improved by the development of sensitive new assays based on real-time monitoring of amplification products (Kami 2001; Costa *et al.*, 2002; Loeffler *et al.*, 2002). Potential advantages of real-time PCR for detection of fungal agents are its speed, convenience and accuracy. Furthermore, the quantitative data provided assists in identification of false positives and may be useful clinically. Real-time PCR technology combines rapid *in vitro* amplification of DNA with real time detection and quantification of the target molecules present in a sample. For example, assays based on the LightCycler®, which include identification of the product and quantification, are completed in approximately 45 minutes (Schnerr *et al.*, 2001). Real-time PCR appears to be highly sensitive for the diagnosis of invasive aspergillosis (IA) and quantitation has been found to be accurate (Ohgoe *et al.*, 1997; Kawamura *et al.*, 1999; Latge, 1999; Kami 2001). The sensitivities of real-time PCR, EIA and the BDG assay were 79, 58 and 67%, respectively, with specificities of 92, 97 and 84%, respectively. The real-time assays should assist early diagnosis and the selection of optimal treatment regimens, so reducing the mortality and morbidity associated with invasive fungal infection.

Sample Preparation

Contamination Control

Contamination of PCR reactions with fungal spores and both fungal and bacterial DNA is a recognized problem (Loeffler *et al.*, 1999; Corless *et al.*, 2000). When primers that target conserved sequences are used residual fungal DNA in the reagents used for extraction may lead to significant background problems. Particular caution is required in selecting enzymes derived from fungal sources (Loeffler *et al.*, 1999). It is recommended that separate rooms be used for extraction of fungal DNA, preparation of PCR reagents, setting-up the PCR reactions and for handling the PCR amplicons. The use of aerosol barrier tips and the use of one-way laboratory coats, disposable gloves and masks can also reduce the incidence of contamination. Methods of eliminating amplicons from the environment are also helpful, for example, by UV illumination either with or without 8-methoxypsoralen pre-treatment. Alternatively, amplicons may be eliminated from the PCR reaction before cycling by using the uracil-DNA-glycosylase system. Fungal cultures should be handled in exhaust protective cabinets to avoid the dissemination of spores in the laboratory. It is essential to include multiple negative controls in each run and it has been suggested that one control for every five samples is appropriate (Loeffler *et al.*, 1999).

Extraction of Fungal DNA

There are several different approaches to the extraction of fungal DNA from blood. Most workers start by removing red cells and then extract the white cell pellet. However, fungal DNA can also be extracted from serum, plasma or environmental specimens (Costa *et al.*, 2001; Haugland *et al.*, 2002). We have used proteinase K and lyticase to effect lysis and QIAmp columns for DNA purification (Qiagen, Crawley, UK). The protocol is included later in this chapter.

Target Sequences, Probes and Primers for Invasive Fungal Infection

Although some workers have used single copy genes as targets for fungal pathogens most attention has focused on conserved multi-copy genes, such as ribosomal RNAs, where it is possible to select primers that amplify all the major fungal species associated with invasive disease. Sufficient variation between rRNA sequences is present such that the organism may be identified to the species level from the internal sequence of the PCR amplicon. Either of the two classes of rRNA present in fungal cells, the nuclear 18S or the mitochondrial 16S may be targeted. The mitochondrial gene is present in greater abundance (*i.e.* one per mitochondrion) but there may be specificity problems associated with using this gene.

The primers shown in Table 1 bind to regions of the 18S rRNA gene, which are highly conserved throughout the whole fungal kingdom and enable the detection of all medically relevant fungal pathogens. The amplicons produced are between 490 bp and 504 bp in size, depending on the fungal species and include a variable region. Hybridization probes that detect the clinically most relevant fungal pathogens, *A. fumigatus* and *C. albicans* have been described (Loeffler *et al.*, 2000). The *A. fumigatus* probe is not species-specific but the other fungal species that match perfectly, including several other *Aspergillus* species are not known human pathogens. The probes hybridise to the *A. fumigatus* and *C. albicans* amplicons at adjacent sites, separated

Table 1. Primers and probes for fungal real-time PCR.

Oligonucleotide	Specificity	Sequence (5' to 3')
Forward primer (18SF)	Fungal	ATTGGAGGGCAAGTCTGGTG
Reverse primer (18SR)	Fungal	CCGATCCCTAGTCGGCATAG
Hybridization probe acceptor	*A. fumigatus* *C. albicans*	**LC**-TGAGGTTCCCCAGAAGGAAAGGTCCAGC **LC**-TGGCGAACCAGGACTTTTACTTTGA
Hybridization probe donor	*A. fumigatus* *C. albicans*	GTTCCCCCCACAGCCAGTGAAGGC -**F** AGCCTTTCCTTCTGGGTAGCCATT –**F**

LightCycler Red 640 (**LC**) and fluorescein (**F**).

by two bases. On hybridisation the fluorescein and LightCycler Red 640 fluors are brought into close proximity and the efficiency of FRET increases.

Standards for Fungal Real-time PCR

A. fumigatus conidia (Figure 1) grown on Sabouraud Dextrose Agar (SDA) can be used as an external standard for real-time PCR. Conidia are counted and then spiked into uninfected blood to provide useful extraction and PCR efficiency controls.

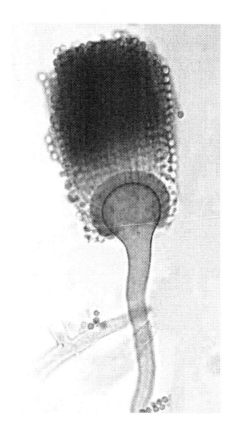

Figure 1. The micrograph shows an *Aspergillus fumigatus* conidial head. The vesicle is approximately 40 microns in diameter.

Protocols

Preparation of *A. fumigatus* Standards

Aspergillus fumigatus was cultured on SDA for 72 h at 30°C. Aspergillus colonies were quantified by hemocytometry and diluted to 10^1-10^6 CFU/ml and then spiked into blood samples from healthy volunteers for use as a quantified standard. Colonies grown in this way are also suitable for the preparation of a naked DNA positive control as described below.

Fungal Nucleic Acid Extraction from Blood

All solutions should be prepared using molecular biology grade reagents and plastics, before dividing into aliquots of suitable size ready for use. Proteinase K stock (Roche Diagnostics, Mannheim; 20 mg/ml) may be stored at –20°C. Defrosted protease K is stable for at least 2 months when stored at 2-8°C and should not be subjected to multiple freeze/thaw cycles All manipulations should be performed in class II cabinets to reduce the possibility of contamination.

Mix EDTA treated blood (1 ml) with 5 ml RCLB (red cell lysis buffer containing 10 mM Tris-HCl pH 7.6, 5 mM $MgCl_2$ and 10 mM NaCl) and incubate on ice for 10-15 minutes. Centrifuge the lysate (3000g for 10 minutes) in a sealed rotor bench centrifuge, discard the supernatant and then add fresh RCLB to the pellet, incubate and recentrifuge to collect the cell pellet. Resuspend the pellet in 1 ml WCLB (white cell lysis buffer is RCLB with proteinase K stock solution added to a final concentration of 200 μg/ml) in a microfuge tube and incubate at 65°C for 45 minutes. Centrifuge at 3000g for 10 minutes and then discard the supernatant. Resuspend the pellet in 50 mM NaOH (200 μl) and then cover with a drop of mineral oil and incubate at 95°C for 10 minutes. Fungal material is recovered by centrifugation (3000g, 10 minutes), resuspended in lyticase solution (0.5 ml containing 300 μg/ml lyticase [Sigma, L-2524], 50 mM Tris, pH 7.5, 10 mM EDTA, and 28mM β-mercaptoethanol) and incubated at 37°C for 30 minutes to produce spheroplasts. The spheroplasts are harvested by centrifugation (10,000g in a sealed rotor microcentrifuge for 10 minutes). Cell lysis in ATL buffer plus proteinase K (180 µl ATL and 20 µl stock proteinase K) and DNA extraction is done using the QIAamp DNA mini kit (Qiagen, Crawley, UK) reagents and tissue protocol. The eluted nucleic acid is stored at -70°C.

Real-Time PCR

Two examples of PCR protocols are given below:

SYBR Green I Method

The reaction components for the SYBR Green I assay for the LightCycler™ is shown in Table 2. Nine µl aliquots of this mix are transferred into separate tubes or wells and 1 µl of sample DNA is added. Many alternative formulations will give equivalent results. For example, the BioGene master mix and SYBR Green I can be replaced by the Roche LightCycler™ DNA Master SYBR Green I mix with appropriate volume adjustments.

The total volume (*i.e.* 10 µl) is carefully pipetted into a LightCycler capillary and the lid is applied loosely. The reaction mixture is spun to the bottom of the capillary by gentle centrifugation (2000 rpm for

Table 2. Mix for 10 µl LightCycler™ reactions using the SYBR Green I method.

Reagent	Volume (µl)
Master mix (BioGene, Kimbolton)*	5.0
H₂O (Sigma, Poole; PCR grade)	2.5
Primer mix (5 µM)	1.0
SYBR Green I (BioGene, Kimbolton; 1:1000)	0.5
Total volume	9

*These mixes contain buffer salts, stabilizer and polymerase. Versions are available containing different concentrations of magnesium ions. For the fungal 18S rRNA assay the final magnesium concentration was 3mM.

5 second) and then the lids are sealed. The capillaries are loaded into the LightCycler and run under the following conditions:

- Initial denaturation for 1 min at 95°C.
- Forty amplification cycles comprised of the following steps: denaturation at 95°C for 0s, annealing at 60°C for 5s and extension at 72°C for 10s with fluorescence acquisition in the F1 channel (530 nm). Maximum ramp rates are used throughout.
- Melting curve analysis follows amplification directly and is comprised of a maximum rate temperature transition to 45°C held for 5s and then a transition to 95°C at a rate of 0.2°C/s with continuous monitoring of fluorescence in the F1 channel.
- Finally the samples are cooled to 40°C prior to removal from the instrument.

The melt temperatures for the fungal 18S rRNA PCR products should be in the range 84-86°C. If required each reaction can be made in a total volume of 20µl providing the concentrations of reagents are maintained.

Hybridisation Probe Method

The master mix for detection of the fungal 18S gene using hybridization probes is shown in Table 3. Equal volumes of master mix and template are added to each capillary as described above. The amplification

Table 3. The master mix for detection of the fungal 18S gene using hybridization probes.

Reagent	Volume (µl)	Final
Roche LightCycler FastStart DNA Master Hybridization Probes	2	1X
$MgCl_2$ (25 mM)	1.6	3 mM
Primers (12 µM each)	1	0.6 µM each
Probes (30 µM each)	2	3 µM each
H_2O (PCR grade)	3.4	
Total volume	10	

program was as follows:

- Denaturation for 9 min at 95°C. This step is necessary to activate the hot start *Taq* polymerase.
- Forty-five amplification cycles comprised of the following steps: denaturation at 95°C for 1s, annealing at 54°C for 15s with fluorescence acquisition in the F2 channel (640 nm) and finally extension at 72°C for 25s. Maximum ramp rates are used throughout.
- Finally the samples are cooled to 40°C prior to removal from the instrument.

Separate reactions are run for detection of either *A. fumigatus* and *C. albicans* containing the respective probes. It is not necessary to determine the melting temperature of the probe/amplicon complex to confirm that the expected product has been formed due to the specificity of the probe reactions.

Analysis for Quantification

For each run in which quantification is required extracts of the *A. fumigatus* standards should be included run in parallel reaction tubes. A standard curve of crossing threshold cycle number against log CFU is plotted and used to in the calculation of the number of copies of the target sequence in the samples.

References

Anaissie, E. 1992. Opportunistic mycoses in the immunocompromised host: experience at a cancer center and review. Clin. Infect. Dis. 14 Suppl 1: S43-53.

Becker, M.J., de Marie, S., Willemse, D., Verbrugh, H.A., and Bakker-Woudenberg, I.A. 2000. Quantitative galactomannan detection is superior to PCR in diagnosing and monitoring invasive pulmonary aspergillosis in an experimental rat model. J. Clin. Microbiol. 38: 1434-1438.

Corless, C.E., Guiver, M., Borrow, R., Edwards-Jones, V., Kaczmarski, E.B., and Fox, A.J. 2000. Contamination and sensitivity issues with a Real-time universal 16S rRNA PCR. J. Clin. Microbiol. 38: 1747-1752.

Costa, C., Vidaud, D., Olivi, M., Bart-Delabesse, E., Vidaud, M., and Bretagne, S. 2001. Development of two Real-time quantitative TaqMan PCR assays to detect circulating Aspergillus fumigatus DNA in serum. J. Microbiol. Methods 44: 263-269.

Costa, C., Costa, J.M., Desterke, C., Botterel, F., Cordonnier, C., and Bretagne, S. 2002. Real-time PCR coupled with automated DNA extraction and detection of galactomannan antigen in serum by enzyme-linked immunosorbent assay for diagnosis of invasive aspergillosis. J. Clin. Microbiol. 40: 2224-2227.

Denning, D.W. 1998. Invasive aspergillosis. Clin. Infect. Dis. 26: 781-805.

Einsele, H., Hebart, H., Roller, G., Loffler, J., Rothenhofer, I., Muller, C.A., Bowden, R.A., van Burik, J., Engelhard, D., Kanz, L., and Schumacher, U. 1997. Detection and identification of fungal pathogens in blood by using molecular probes. J. Clin. Microbiol. 35: 1353-1360.

Haugland, R.A., Brinkman, N., and Vesper, S.J. 2002. Evaluation of rapid DNA extraction methods for the quantitative detection of fungi using Real-time PCR analysis. J. Microbiol. Methods 50: 319-323.

Kami, M. 2001. Current approaches to diagnose invasive aspergillosis: application of Real-time automated polymerase chain reaction. Nippon Ishinkin Gakkai Zasshi. 42: 181-188.

Kami, M., Fukui, T., Ogawa, S., Kazuyama, Y., Machida, U., Tanaka, Y., Kanda, Y., Kashima, T., Yamazaki, Y., Hamaki, T., Mori, S., Akiyama, H., Mutou, Y., Sakamaki, H., Osumi, K., Kimura, S., and Hirai, H. 2001. Use of Real-time PCR on blood samples for diagnosis of invasive aspergillosis. Clin. Infect. Dis. 33: 1504-1512.

Kawamura, S., Maesaki, S., Noda, T., Hirakata, Y., Tomono, K., Tashiro, T., and Kohno, S. 1999. Comparison between PCR and detection of antigen in sera for diagnosis of pulmonary aspergillosis.

J. Clin. Microbiol. 37: 218-220.

Latge, J.P. 1999. Aspergillus fumigatus and aspergillosis. Clin. Microbiol. Rev. 12: 310-350.

Loeffler, J., Hebart, H., Bialek R., Hagmeyer L., Schmidt D., Serey FP., Hartmann M., Eucker J., and Einsele H. 1999. Contaminations occurring in fungal PCR assays. J. Clin. Microbiol. 37: 1200-1202.

Loeffler J., Henke N., Hebart H., Schmidt D., Hagmeyer L., Schumacher U., and Einsele H. 2000. Quantification of fungal DNA by using fluorescence resonance energy transfer and the light cycler system. J. Clin. Microbiol. 38: 586-590.

Loeffler, J., Hebart, H., Cox, P., Flues, N., Schumacher, U., and Einsele, H. 2001. Nucleic acid sequence-based amplification of Aspergillus RNA in blood samples. J. Clin. Microbiol. 39: 1626-1629.

Loeffler, J., Schmidt, K., Hebart, H., Schumacher, U., and Einsele, H. 2002. Automated extraction of genomic DNA from medically important yeast species and filamentous fungi by using the MagNA Pure LC system. J. Clin. Microbiol. 40: 2240-2243.

Ohgoe, K., Miyanishi, S., Aihara, M., and Matsuo, S. 1997. Significance of Aspergillus fumigatus rDNA detected by polymerase chain reaction in diagnosis of pulmonary aspergillosis] Kansenshogaku Zasshi. 71: 507-512.

Schnerr, H., Niessen, L., and Vogel, R.F. 2001. Real time detection of the tri5 gene in Fusarium species by lightcycler-PCR using SYBR Green I for continuous fluorescence monitoring. Int. J. Food Microbiol. 71: 53-61.

Skladny, H., Buchheidt, D., Baust, C., Krieg-Schneider, F., Seifarth, W., Leib-Mosch, C., and Hehlmann, R. 1999. Specific detection of Aspergillus species in blood and bronchoalveolar lavage samples of immunocompromised patients by two-step PCR. J. Clin. Microbiol. 37: 3865-3871.

Van Burik, J.A., Myerson, D., Schreckhise, R.W., and Bowden, R.A. 1998. Panfungal PCR assay for detection of fungal infection in human blood specimens. J. Clin. Microbiol. 36: 1169-1175.

Index

Index

Index

HIV 51, 68, 126, 176, 226, 233-236, 241, 244-245, 271, 287, 308
HIV-1 *see* HIV
HK *see* housekeeping gene
*hly*A gene 251
hMPV *see* human metapneumovirus
hook effect 40, 56
hot-start 40, 43, 73-74, 79, 94, 218
housekeeping gene 89, 116, 160-161, 164-166, 181, 182
HPLC *see* high performance liquid chromatography
HSV 1 *see* herpes simplex virus
HSV 2 *see* herpes simplex virus
HSV *see* herpes simplex virus
human adenovirus 293, 313
human amniotic fluids 273-274
human cytomegalovirus *see* cytomegalovirus
human hepatitis B *see* hepatitis B virus
human herpesvirus 283, 324
human herpesvirus 6 277, 309, 312, 314, 317
human herpesvirus 7 278
human herpesvirus 8 278, 309, 323
human metapneumovirus 275, 278, 281-282, 310, 317
Hybeacon 57-58, 65
hybridisation probe 4, 19, 21, 22, 40, 54-58, 75, 80-82, 90, 92, 96, 117, 153, 185-191, 193, 198-202, 206, 257, 265, 270-271, 296, 299, 324, 335
hydrolysis probe 4, 7, 18-22, 24, 39, 48, 50-54, 59-61, 63, 66, 68-69, 75, 77, 79-80, 88, 92-94, 96, 117, 121, 148-149, 153, 158, 174, 177, 185-186, 191-193, 195, 198-199, 255-256, 264-265, 268, 270-274, 279, 281-288, 290-294, 298, 301, 303, 308-309, 312-313, 315-316, 319-320, 323, 337
IA *see* invasive aspergillosis
IC *see* internal control
iCycler IQ 16-17, 22-23, 48, 190, 196
Idaho Technology RAZOR 26
immediate-early antigen gene 276
ImProm-II RT 134
influenza virus A 180, 279
influenza virus B 279
influenza virus 275, 279
inhibition controls 80-81, 107, 109, 111, 284, 307, 309
inhibition 33, 81, 85, 101, 107, 115, 117, 138, 140, 172, 174, 176, 183-184, 288-289, 292, 305
inhibitors 79-80, 86, 105, 107-108, 112, 115, 128-129, 136, 138-139, 158, 164, 169-170, 173, 180, 230, 235, 257, 279
insertion sequence 86, 90, 186, 250, 255-256, 262, 310, 322
inter-assay variation 136
intercalating dye 4, 190, 196, 197
internal control 5, 56, 85-90, 92, 98-101, 108-109, 117, 141, 159, 160, 164, 174, 176, 181-182, 184, 235, 257, 284, 287-288, 302, 305, 319

inter-run variation 109
IntraTaq 63
invasive aspergillosis 329, 337
invasive fungal infections 327-329
IS1001 *see* insertion sequence
IS481 *see* insertion sequence
IS6110 *see* insertion sequence
isoniazid resistance 200-201, 206, 209, 295-297, 312, 324
isothermal 8, 229, 230, 231, 243, 244
Ixodes scapularis ticks 264, 325
JOE *see* carboxy-4', 5'-dichloro-2', 7'-dimethoxyfluorescein
junin 285
K5 gene 278
katG gene 201, 295, 297
kinetic PCR 14
kinetoplast 269
L gene 278, 282
L. brasiliensis 269, 274-275
L. pnuemophila see Legionella pnuemophila
lamivudine 11, 212, 214, 220-222, 225-226, 291
LANA *see* latency-associated nuclear antigen
large tegument protein gene 278
laser assisted microdissection 11, 121-122
laser capture microdissection 166
Lassa virus 276, 285, 310
LASV *see* Lassa virus
latency-associated nuclear antigen 235
LC Red 640 49, 55, 189, 202, 331, 332
LC Red 705 49, 55, 189, 202
LCGreen I 11, 196, 210
LCM *see* laser capture microdissection
LED *see* light emitting diode
Lef gene 250, 299
Legionella bozemanii 257
Legionella dumoffii 257
Legionella fairfieldensis 260
Legionella longbeachae 257
Legionella micdadei 257
Legionella pnuemophila 257-260, 308, 322. 325-326
Legionella worsleinsis 260
Legionnaires disease 257
Leishmania brasiliensis 269, 274-275
Leishmania donovani 269, 274-275
Leishmania infantum 269, 309
Leishmania species 269, 274-275, 319, 332
Leishmaniasis 274, 326
light emitting diode 17
LightCycler 5,11, 15, 17, 19, 23-25, 28 36, 38-40, 48-50, 53-56, 70, 73, 78, 80, 83, 107, 117-119, 122, 177, 186-187, 189-190, 196, 198-203, 205, 207-208, 210, 217, 249, 300, 308-309, 311, 313-319, 321, 323-325, 329, 334-335, 338
Light-up probe 57-59, 66, 69, 76
limit of detection 17, 241, 257, 267, 272, 276, 284, 286, 291, 298, 302-303
liver damage 220, 292

Index

344

Printed in the United Kingdom
by Lightning Source UK Ltd.
106167UKS00001B/40-222